ECE/STEEL/105

ECONOMIC COMMISSION FOR EUROPE
Geneva

Iron and steel scrap: its significance and influence on further developments in the iron and steel industries

Seventh updating

UNITED NATIONS
New York and Geneva, 1997

NOTE

The designations employed and the presentation of the material in this publication do not imply the expression of any opinion whatsoever on the part of the Secretariat of the United Nations concerning the legal status of any country, territory, city or area, or of its authorities, or concerning the delimitation of its frontiers or boundaries.

ECE/STEEL/105

UNITED NATIONS PUBLICATION

Sales No. E.97.II.E.15

ISBN 92-1-116672-1
ISSN 1014-9163

TABLE OF CONTENTS

TABLES

FIGURES

SYMBOLS EMPLOYED

...	=	figure not available
-	=	nil or negligible quantity
Mt	=	millions of metric tonnes

PREFATORY NOTE

The secretariat of the United Nations Economic Commission for Europe (UN/ECE) has been publishing biennial reviews on iron and steel scrap since 1983. The present review was prepared by Jacques Astier and Jeong-Shik Kwak.

The study basically consists of statistical tables, but short texts have been included to introduce certain sets of figures so as to make them easier to understand or to indicate the main trends which they reflect, in so far as this is possible on the basis of information covering a short period.

In some cases, where information for the review was available from a well-established source other than the secretariat questionnaire, figures have been supplemented or updated in such a way as to keep them as homogeneous and continuous as possible.

The present review consists of eight chapters, dealing with steel production, scrap supply and demand, scrap trade and prices, and future trends. Annex 1 provides the total scrap balance in tabular form for each of the 51 countries that supplied data in response to the ECE questionnaire. Annex 2 has 27 statistical tables covering production, consumption by processes, requirements and trade.

DEFINITIONS USED IN THIS STUDY

The **iron and steel industry** manufactures primary iron and steel products consisting of all processes from smelting in blast furnaces to the semi-finished stage in rolling mills and founding. That is the production of billets, blooms, slabs or bars; hot and cold rolling and drawing into basic forms such as sheets, tin plate, terne plate and black plate, strips, tubes and pipes, rails, rods; wire rods and heavy gauge wires, castings and forgings.

Scrap includes alloy iron and steel scrap and cast iron scrap.

Circulating scrap comprises any iron or steel arising during the production of iron or steel which is suitable for remelting; runners and other steel-pouring scrap (normal or bottom-poured), including funnels and gates, waste from delivery pipes in bottom pouring, etc., and also rejected and defective ingots not included in production. Circulating scrap also includes, in addition to all hot-rolled production, scrap collected during the production of cold-rolled narrow strip, electrical sheet and strip, tinplate and tin-free steel, sheet and strip with metallic coating, sheet and strip with non-metallic coating, bright bars, drawn wire, seamless tubes including hollow profiles, welded tubes including hollow profiles, tube fittings, and cold-formed sections.

Pig iron production refers to liquid hot metal produced mainly in blast furnaces, but sometimes also in new smelting reduction processes like Corex.

Direct reduction processes refer to the reduction of iron ore in its solid state usually at temperatures below 1,000°C through processes like Midrex or Hylsa (HYL). The product is called pre-reduced metal, DRI (direct reduced iron) or HBI (hot briquetted iron).

Prompt scrap is produced during the transformation or the processing of steel into finished products such as automobiles, cans and metal structures.

Obsolete scrap originates from obsolete finished products which are at the end of their useful life and **capital scrap** is derived from the dismantling of industrial structures such as industrial buildings or machines.

Shredded scrap is obtained by the crushing of light scrap such as car bodies or home appliances like refrigerators.

Independent steel and iron foundries are all steel and iron foundries which are not integrated in an iron and steel plant as defined above.

Flowchart for steel production and scrap collection

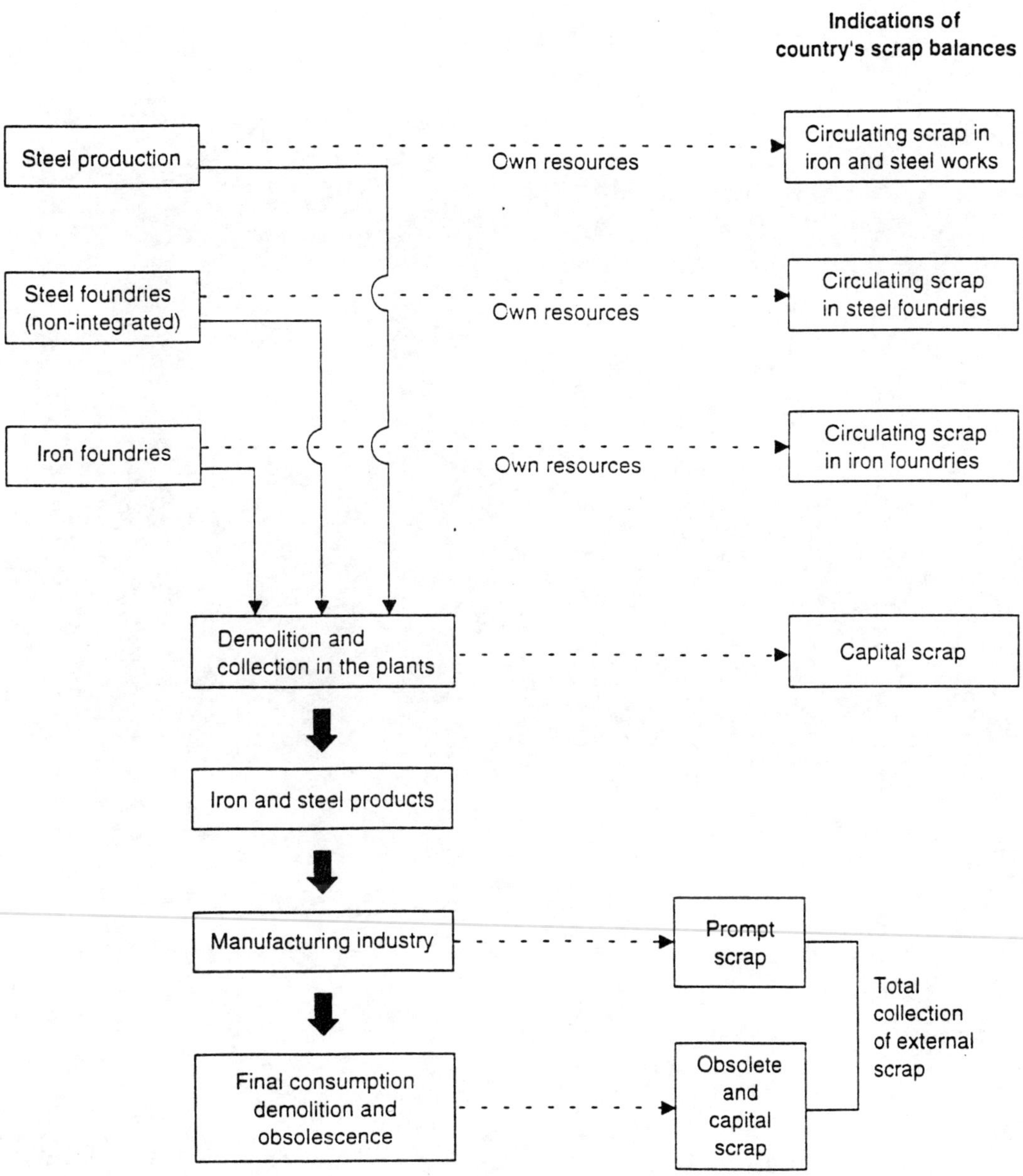

CHAPTER 1

TRENDS IN WORLD STEEL PRODUCTION

The most important factors affecting both the scrap collection and utilization are:

- the evolution of steel consumption and production worldwide;
- the evolution of the structure of steel production plants;
- the impact of new technologies in steelmaking;
- the impact of new developments in continuous casting;
- the production of primary metals, pig iron (or, in liquid form, hot metal), DRI/HBI and iron carbide.

These developments will be examined in the first five sections of the present chapter. The sixth and final section gives a quantitative view of the world steel industry, emphasizing the role of scrap, or more exactly, the various categories of scrap as defined in the introduction to this study.

1.1 Evolution of steel consumption worldwide

On this subject, reference will be made to the analysis of the steel market made by UN/ECE:

(a) *The Steel Market in 1995 and Prospects for 1996* (ECE/STEEL/98)

"The steel market continued to improve through 1994 into the first half of 1995. However, it peaked before the summer. In the latter half of the year, it weakened because of increasing inventories. World apparent consumption of finished steel increased by 0.2 per cent to 633 Mt in 1994. Total world crude steel production amounted to 748 Mt, an increase of 3.1 per cent or 23 Mt compared to 1994. Extensive new crude steel-making capacity is being added in North America and Asia. New capacity will reach about 18 Mt in North America and 58 Mt in Asia from 1995/96 to 1998/99. The steel market is likely to remain weak in the first half of 1996.

(b) Presentations on the steel market in 1996 and prospects for 1997 made by the UN-ECE secretariat (March 1997):

" World economic growth is expected to pick up further in the course of 1997, although its growth will be modest rather than strong... World crude steel production reached 750.8 Mt in 1996, a slight decrease of 0.2 per cent or 1.2 Mt compared to 1995. All in all, the world steel market is expected to show somewhat favourable trends in 1997. However, there are some concerns ahead. An effort to avoid over-supply and to match supply to the level which the market requires is essential."

(c) In the medium term, IISI has estimated that world consumption of finished steel could reach about 714 Mt in the year 2000.

(d) In the longer term, estimates are, of course, different from one institution to another, but the following trends are apparent, starting from the recent historical data:

Table 1.1 Recent trends and forecast for world steel production and consumption
(Millions of tonnes)

Year	Crude steel production	Finished steel consumption
1990	770	647
1994	725	633
1995	750	652
1996	751	653
2000	800	714
2005	850 [1] 943 [2]	847 [2]
2010	900 [1] 1077 [2]	785 [1] 980 [2]

Source: UN/ECE Working Party on Steel.

[1] Gorham Intertech Conference, Charlotte, North Carolina, United States, 3-5 March 1997 Jacques Astier, Battelle-ITM.

[2] Idem, M. Hecht, Paine Webber.

1.2 Developments in the structure of steel production

The trends indicated in the *Steel Market in 1995* have been accompanied by an important restructuring of integrated iron and steel plants which have become smaller in number except in a few developing countries, and also smaller in unit capacity. There was a corresponding expansion of mini mills, both in number and in unit capacity; this point must be emphasized since a growing number of these so-called "mini mills" now have a capacity of well over 1 Mt/year.

Table 1.2, containing the statistical data for electric steelmaking production versus the corresponding data for both oxygen and open-hearth steel production, shows that the increase in integrated, usually large scale, iron and steel plants was continuous up to the 1980s; and that the number and size of mini mills have grown continuously since the end of the 1970s. Many of them now have a capacity not unlike the so-called "large integrated iron and steel plants".

Table 1.2 Developments in the structure of steel plants

		Plants based on BOF or OH (or others)	Plants based on EAF
Usual range of capacity		1 to 10 Mt/year	under 2 Mt/year
1970	Mt	509	86
	%	86	14
1980	Mt	554	162
	%	77	23
1990	Mt	553	217
	%	72	28
1993	Mt	503	228
	%	69	31
1994	Mt	498	232
	%	68	32
1995	Mt	506	247
	%	67	33
1996	Mt	503	247
	%	67	33

Source: UN/ECE Working Party on Steel and IISI Steel Statistical Yearbook.

1.3 Developments in steelmaking

The trends in steelmaking, worldwide and in specific countries, are shown in figure 1.1 and in the tables in annex 2, particularly in table 2.1.

They are characterized by the following three main features:

(a) A rapid decrease in open-hearth steelmaking, which has completely disappeared in a number of countries: in Japan in 1979 (figure 1.2), in the United States in 1992 (figure 1.3). It is negligible in western Europe (figure 1.4), decreasing very fast in eastern Europe (figure 1.5) but is still important in the Commonwealth of Independent States (figure 1.6) and in a few developing countries, especially China (figure 1.7);

(b) Stagnation in oxygen steelmaking, which is even decreasing in western Europe, the United States and Japan (figures 1.2, 1.3 and 1.4) but increasing in eastern Europe and the CIS (figures 1.5 and 1.6) and in a number of developing countries like China (figure 1.7);

(c) A significant and continuous increase in electric steelmaking (figure 1.1 and 1.10) in all regions of the world.

These three points are very important for the utilization of scrap, as will be shown in detail in chapter 3 (scrap demand) as, unlike the time when the main steelmaking process was open hearth and could use practically any mix of hot metal/pig iron and scrap, the world steel industry is now split between two processes: oxygen steelmaking based on liquid hot metal and using only small amounts of scrap, and electric steelmaking based on scrap (but using, as will be shown later more and more "scrap substitutes").

Despite many attempts, the development of other new steelmaking processes has been very limited. EOF (Energy Optimizing Furnace) is used only in Brazil (at the PAINS plant) and there are a few projects, notably in India.

Figure 1.1 World crude steel production by process

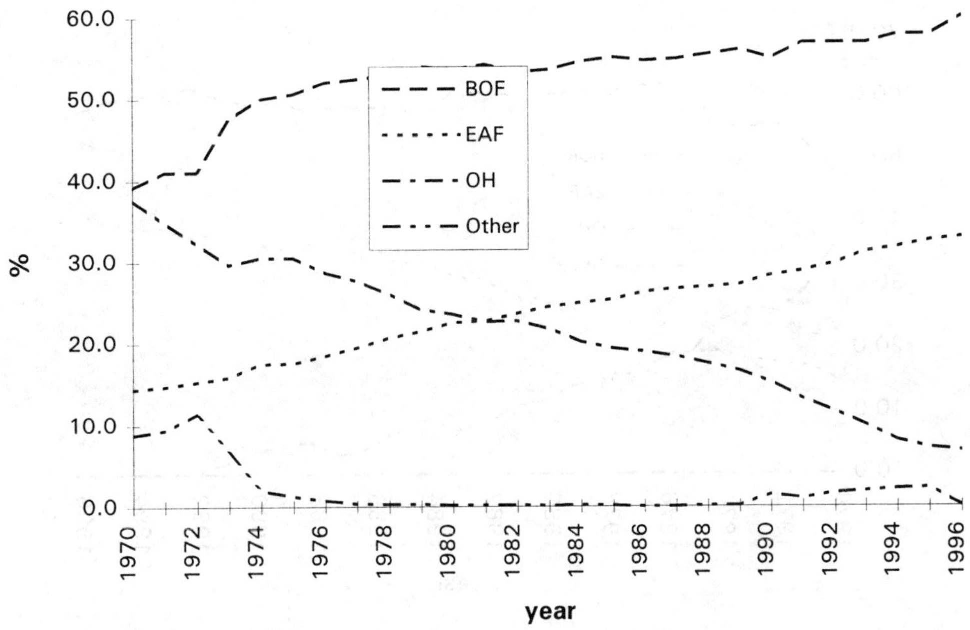

Source: UN/ECE Working Party on Steel.

Figure 1.2 Crude steel production in Japan by process

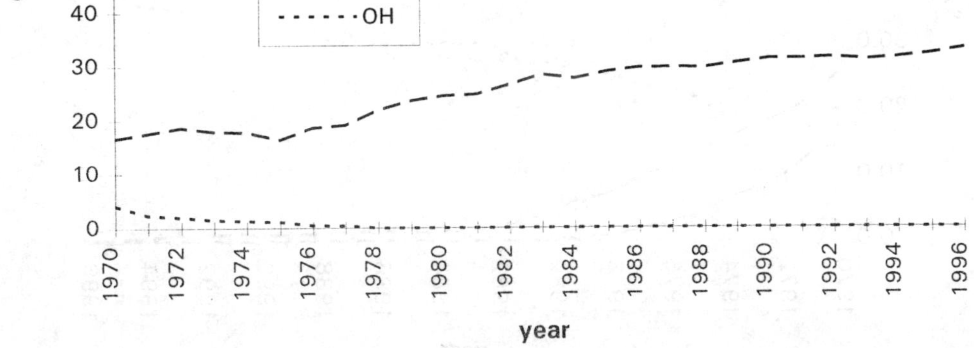

Source: UN/ECE Working Party on Steel.

Figure 1.3 Crude steel production in the United States by process

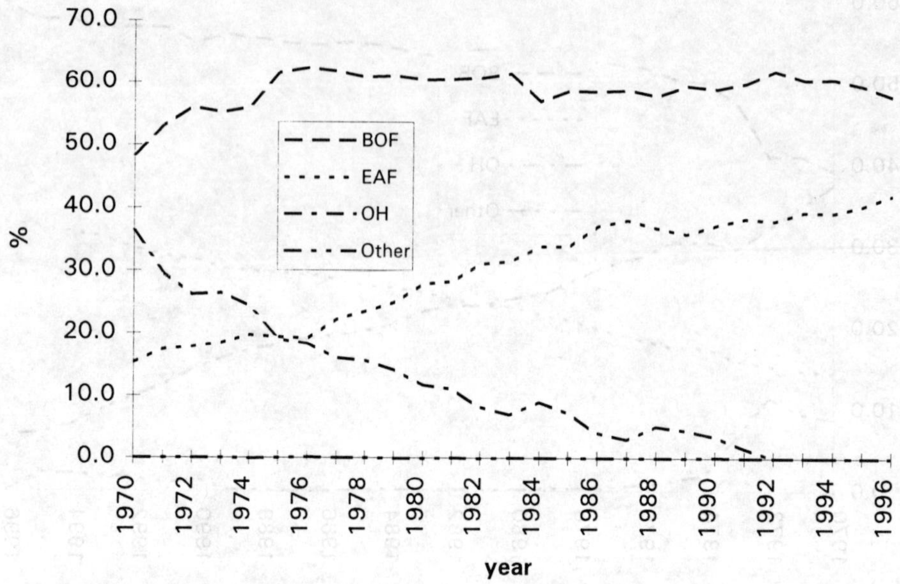

Source: UN/ECE Working Party on Steel.

Figure 1.4 Crude steel production in western Europe by process

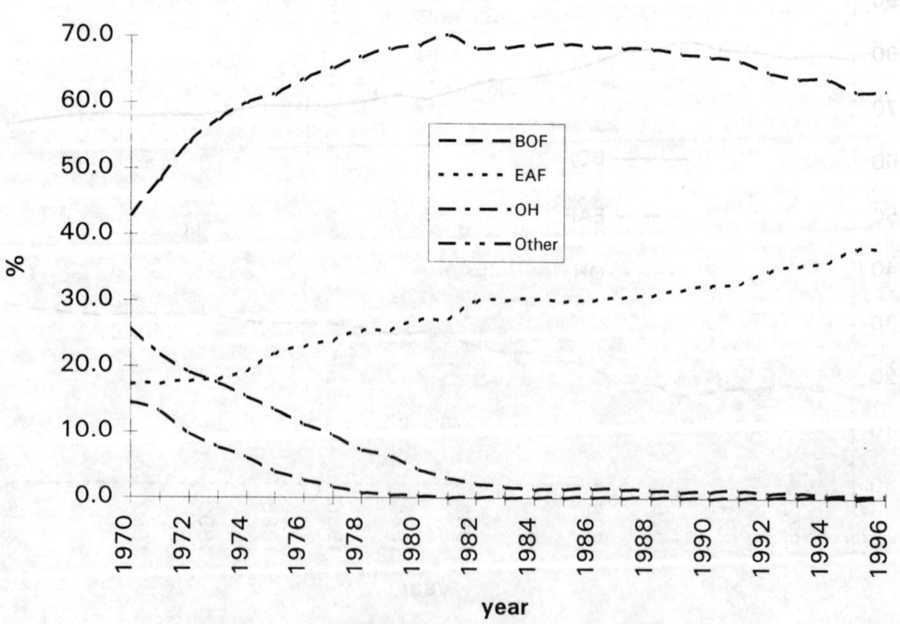

Source: UN/ECE Working Party on Steel.

Figure 1.5 Crude steel production in eastern Europe by process

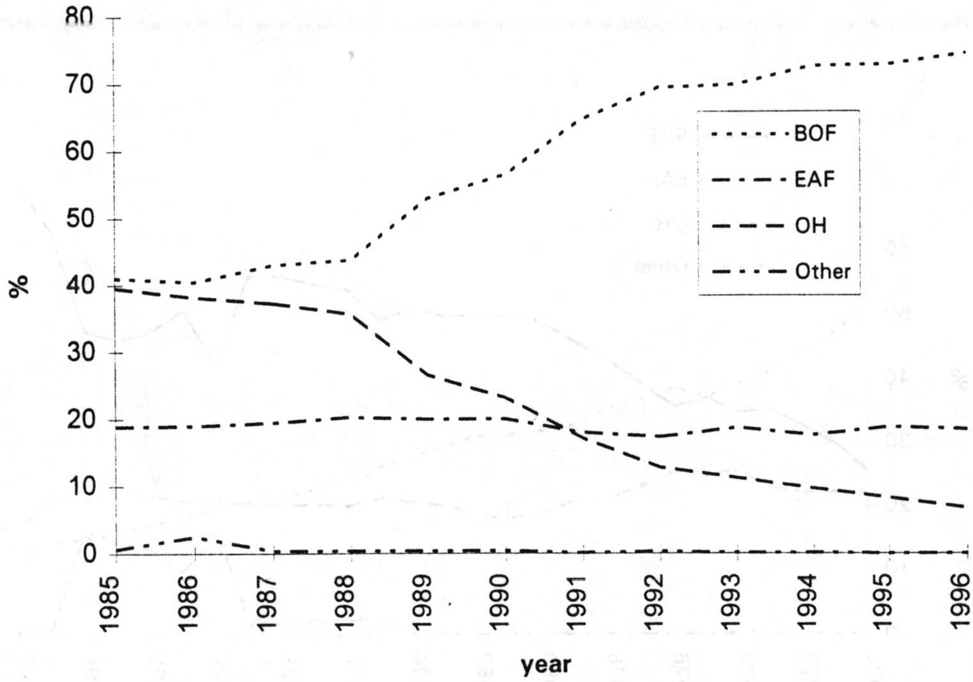

Source: UN/ECE Working Party on Steel.

Figure 1.6 Crude steel production in Commonwealth of Independent States countries by process

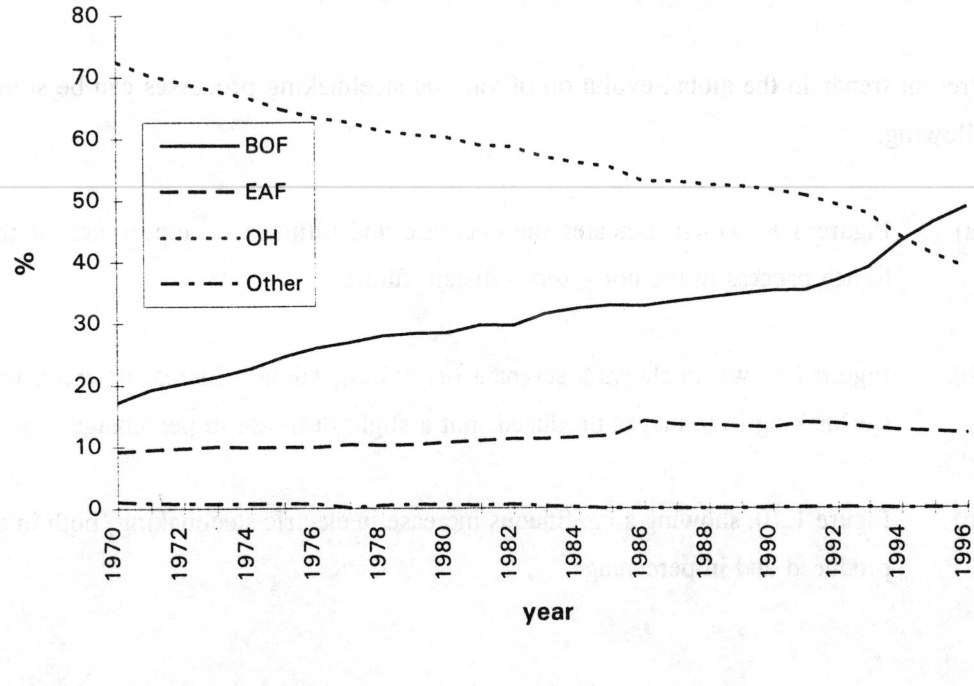

Source: UN/ECE Working Party on Steel.

Figure 1.7 Crude steel production in China by process

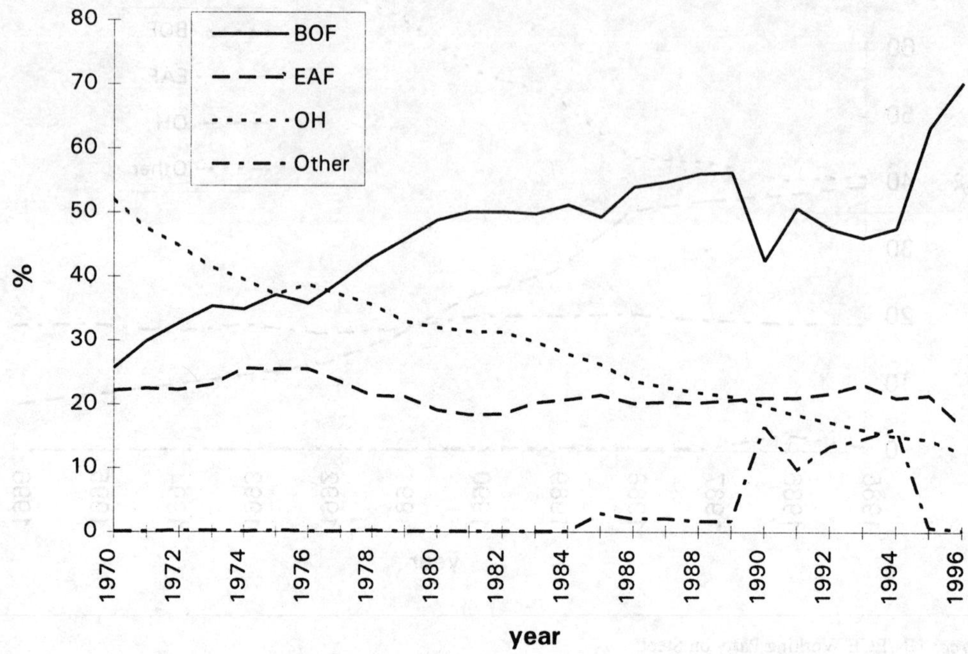

Note: There is still an important production of steel in air blown converters, which could explain the irregularities in BOF and Other during 1985 to 1995.
Source: UN/ECE Working Party on Steel.

Present trends in the global evolution of various steelmaking processes can be summarized by the following:

(a) Figure 1.8, which indicates the decrease and ultimate disappearance of the open-hearth process in the not —too— distant future;

(b) Figure 1.9, which shows a stagnant or, at best, slightly increasing trend in oxygen steelmaking in tonnages produced, but a slight decrease in percentage terms;

(c) Figure 1.10, showing a continuous increase in electric steelmaking, both in tonnages produced and in percentage.

Figure 1.8 Trends in open-hearth steelmaking

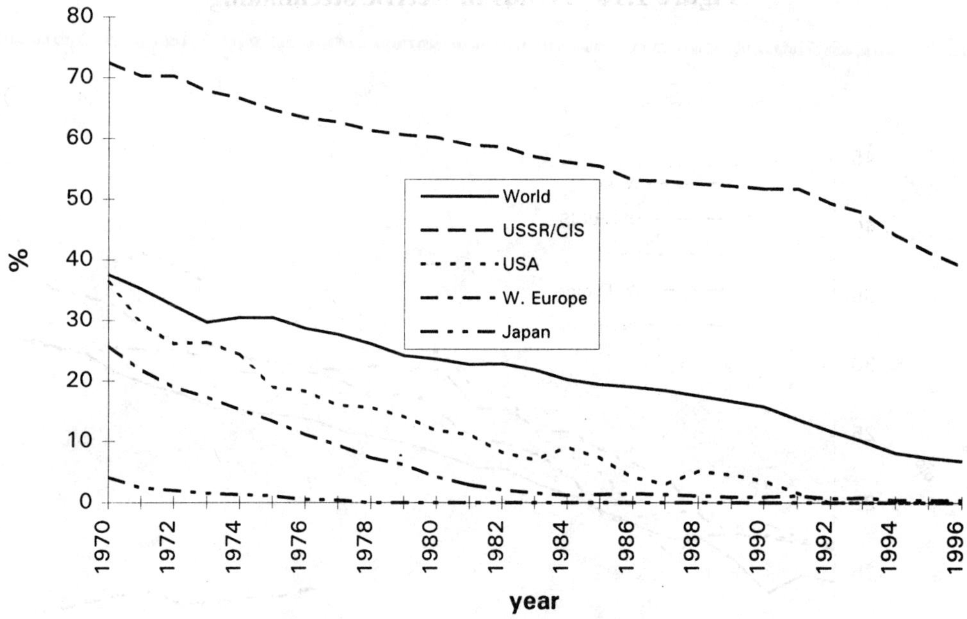

Source: UN/ECE Working Party on Steel.

Figure 1.9 Trends in oxygen steelmaking

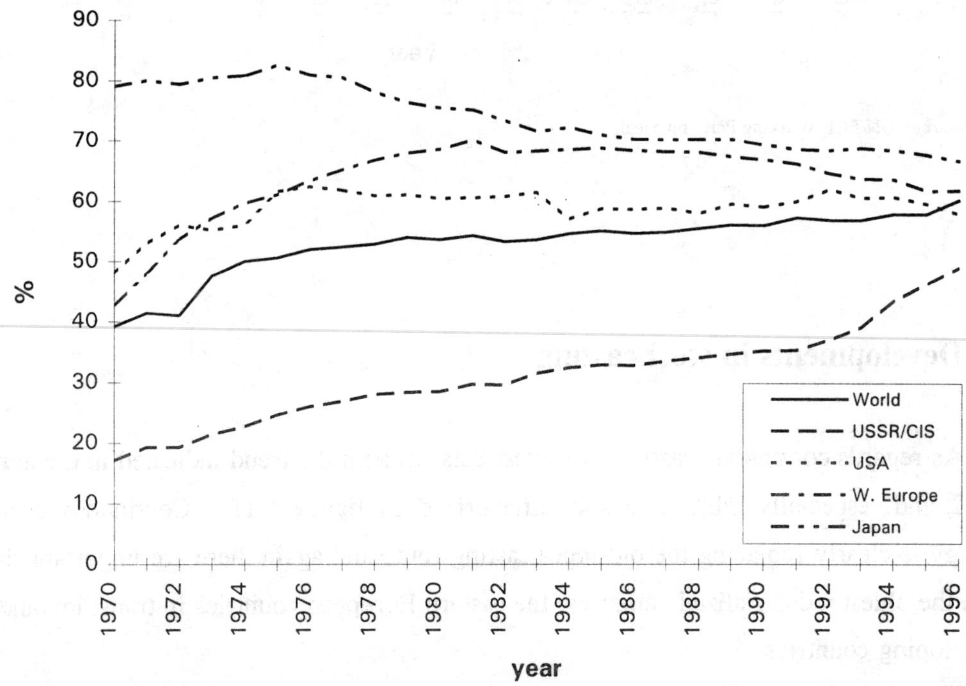

Source: UN/ECE Working Party on Steel.

Figure 1.10 Trends in electric steelmaking

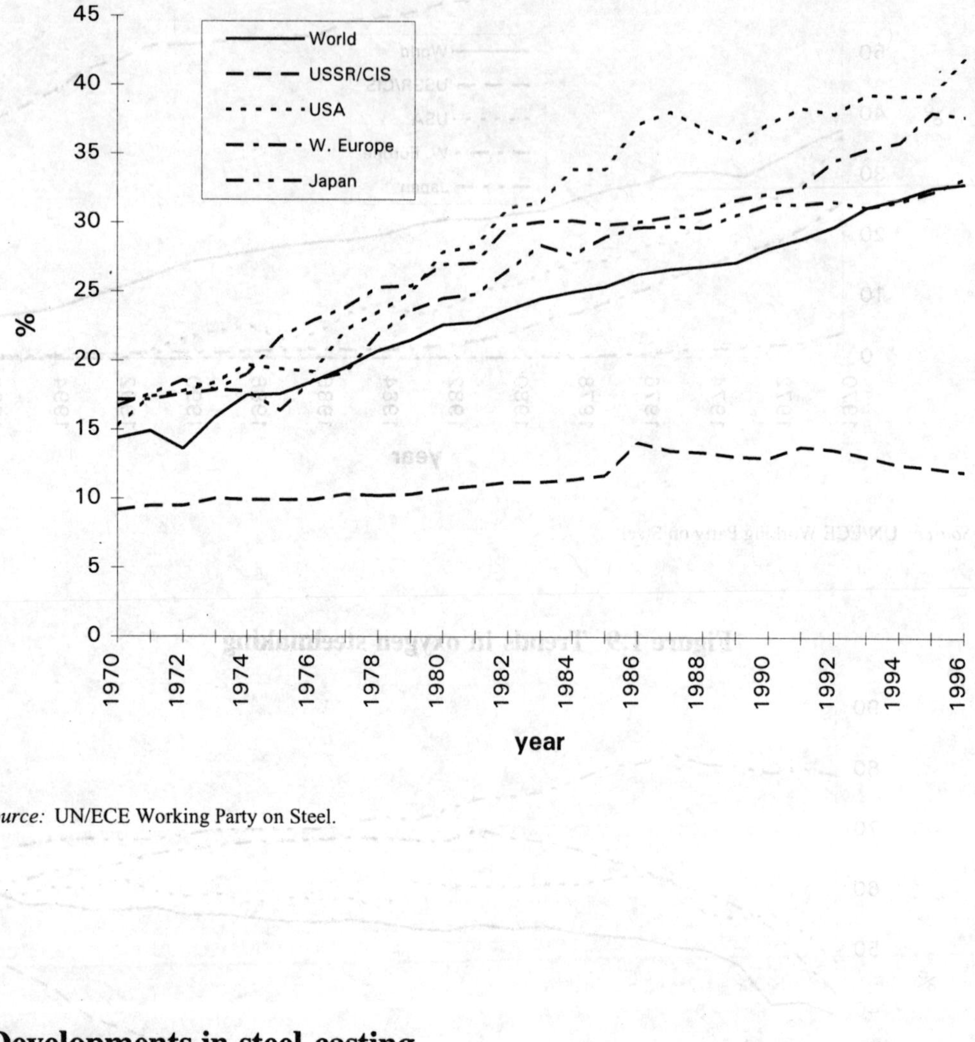

Source: UN/ECE Working Party on Steel.

1.4 Developments in steel casting

As regards continuous casting, the latest data confirm the trend indicated in the annex tables (annex 2 and, especially table 2.2) and summarized in figure 1.11. Continuous casting (CC) technology is clearly replacing the old ingot casting route and again there are important differences between the oldest industrialized countries, the eastern European countries in transition and the CIS; and developing countries.

Figure 1.11 Continuous casting ratio by major regions

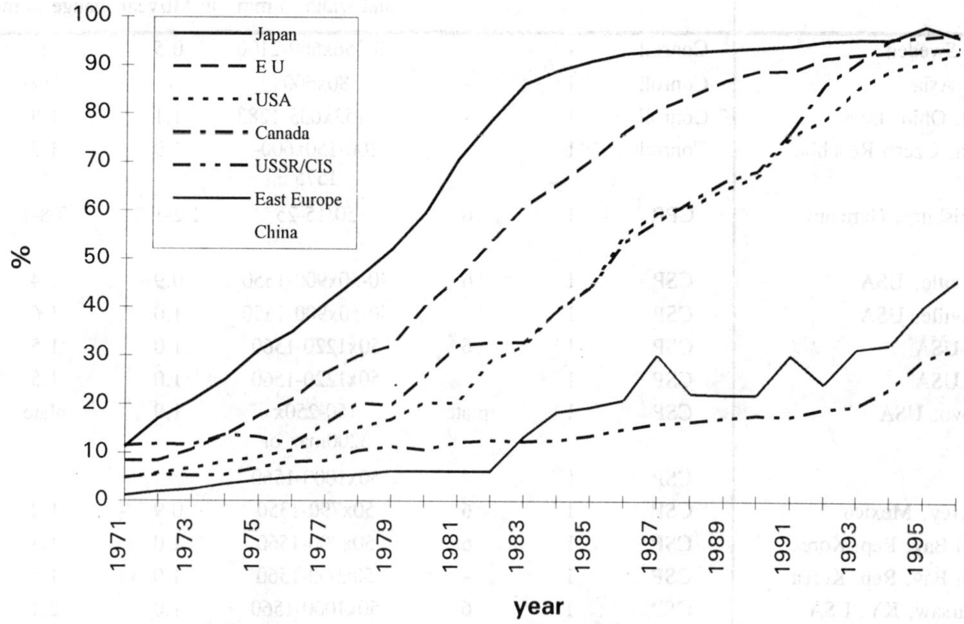

Source: UN/ECE Working Party on Steel.

However, the evolution of continuous casting has to be taken into account. In many countries, conventional or "first generation" CC machines for slabs, blooms and billets have largely (over 90 %) replaced ingot casting but a "second generation"of CC processes is beginning to develop. They belong to the near net shape casting (NNSC) concept for producing semis such as thin slabs and beam blanks.

The development of beam blanks casting (often called "dog bones") has been very fast and this technology is increasingly being used for heavy beams production. However, it is thin slab casting which is developing fastest, especially in the United States but also in Asia as indicated in table 1.3 and table 1.4.

These developments are very important for scrap generation (see later, in chapter 2 on scrap supply) as from ingot casting to "first generation" CC, the amount of "home scrap" in steel plants has decreased from about 0.25 to 0.35 t/t of steel to 0.1 t/t steel. With "second generation" NNSC, this quantity could decrease to 0.05 t/t steel (see figure 1.12).

It should be noted that further developments in NNSC are beginning, especially for stainless steel, as a first step, with strip casting.

Table 1.3 Development of thin slab casting worldwide

Steelworks	Technology	Strands	Mill stands	Cast thickness and width in mm	Capacity in Mt/year	Minimum gauge in mm	Order date	Start-up date
Avesta, Sheffield, Sweden	Conroll	1	-	80-256x660-2100	0.5	2	-	1989
"Edelweiss", S.E. Asia	Conroll	1	-	80x600	-	plates	-	1995
Armco, Mansfield, Ohio, USA	Conroll	1	-	76-152x635-1283	1.1	1.8	Oct.93	Apr.95
Nova Hut, Ostrava, Czech Republic	Conroll	1	-	100-150x600-1575	1.0	1.7		1998
Thyssen Stahl, Duisburg, Germany	CPR	1	6	50/15-25	1.2-1.8	0.8-1	Sep.96	1999
Nucor, Crawfordsville, USA	CSP	1	6	40-50x900-1350	0.9	1.4	Dec.86	Jul.89
Nucor, Crawfordsville, USA	CSP	1		40-50x900-1350	1.0	1.6	May.93	Apr.94
Nucor, Hickman, USA	CSP	1	6	50x1220-1560	1.0	1.5	Nov.90	Aug.92
Nucor, Hickman, USA	CSP	1		50x1220-1560	1.0	1.5	Jun.93	May.94
Geneva Steel, Provo, USA	CSP	1	plate	150-250x 3200max or	1.9	plate	Nov.90	Jul.94
AST, Terni, Italy	CSP	1	-	50x1000-1560	-	-	Sep.91	Dec.92
Hylsa SA, Monterrey, Mexico	CSP	1	6	50x790-1350	0.9	1.2	Dec.92	Feb.95
Hanbo Steel, Asan Bay, Rep. Korea	CSP	1	6	50x900-1560	1.0	1.5	Feb.93	Jun.95
Hanbo Steel, Asan Bay, Rep. Korea	CSP	1	-	50x900-1560	1.0	1.5	Dec.93	Sep.95
Gallatin Steel, Warsaw, KY, USA	CSP	1	6	50x1000-1560	1.0	2.1	Mar.93	Apr.95
Steel Dynamics, Butler, IN, USA	CSP	1	6	40-70x990-1560	1.2	1.2	Mar.94	Dec.95
Ispat Industries, Raigad, Mah, India	CSP	1	6	50x900-1560	1.2	1.2	Mar.94	Aug.96
Aceria Compacta Bizkaya, Bilbao, Spain	CSP	1	6	53x790-1560	0.9	1.3	Jun.94	Jul.96
Acme Metals, Riverdale, IL, USA	CSP	1	7	50x900-1560	0.9	1.25	Jul.94	Oct.96
Nucor, Berkeley, SC, USA	CSP	1	6	50x12220-1680	1.2	1.45	Mar.95	Oct.96
NSM, Chonburi, Thailand	CSP	1	6	45-70x1220-1600	1.2	1.2	Aug.95	Aug.97
Amalgamated Steel Mills, Malaysia	CSP	2	6	50x900-1560	2.0	1.2	Jan.96	Mar.98
Zhujiang Steel, China	CSP	1	5	50x1000-1300	0.8	1.27	Apr.96	Nov.98
Baotou Iron & Steel, China	CSP	2	6	50-70x980-1560	2.0	1.2	Apr.96	-
Handan Iron & Steel, China	CSP	1	1+5	70x900-1680	1.2	1.2	Apr.96	-
Megasteel, Klang, Malaysia	CSP	2	6	50x900-1560	2.0	1.2	Sep.96	Jun.98
Sabolarie project, Buttrio, Italy	FTSR	1	6	50-70x1600	1.0	1.5	-	-
Algoma Steel, Canada	FTSR	2	1R+6F	70x1600	2.0	1.0	Jul.95	Oct.97
Asfahan Steel Co., Iran	FTSR	1	6	60/45x800-1560	0.7	1.5	Jul.96	1999
Jacinto, Villanueva, Philippines	FTSR	1	-	-	1.2	-	Aug.96	Q3/98
Acciaieria ISP di Cremona (Arvedi Group), Cremona, Italy	ISP	1	4	60/43x650-1330	0.7	1.2	May.89	Jan.92
Posco, Kwangyang, Rep. Korea	ISP	2	5	75/60x900-1350	1.8	1.2	Jun.94	Aug.96
Saldanha Steel, Saldanha, South Africa	ISP	2	2R+5F	90/75x900-1560	2.5	1.0	Dec.95	1997
Tuscaloosa Steel, AL, USA	SMS Concast	1	2R+6F	80-100x900-1565	1.0	-	-	Oct.96
North Star/BHP Steel, Delta, OH, USA	Sumitomo/ Danieli	1	2R+6F	80-100x900-1565	1.4	1.0	Oct.94	Nov.96
Trico Steel, Decatur, AL, USA	Sumitomo/ Innse	2	2R+5F	90/70	2.2	1.0	Jun.95	Q1/1997
Ipsco, Muscatine Co., Iowa, USA	TSP	1	Steckel	152x3000	0.9	2.3	Apr.94	Q1/1997
Nova Hut, Czech Republic	TSP	1	single stand reversing	125x800-1600	1.0	1.5	Oct.94	Jun.05

Source: UN/ECE questionnaire and *Steel Times International*, November 1996.

Figure 1.12 Schematic differences between:

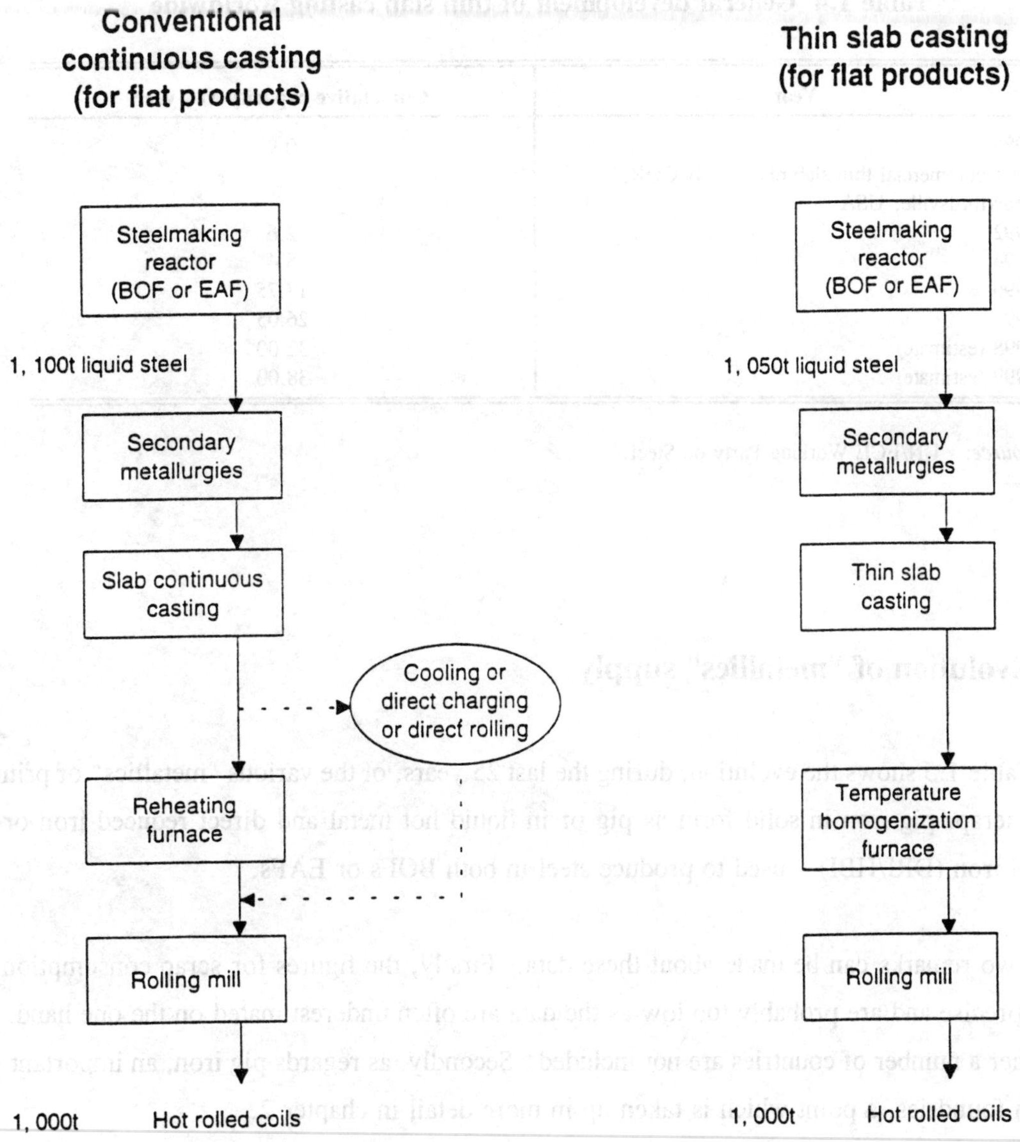

Source: UN/ECE Working Party on Steel.

Table 1.4 General development of thin slab casting worldwide

Year	Cumulative capacity M/t year
1989 First commercial thin slab plant at NUCOR, Crawfordsville, USA	0.9
1992	2.6
1994	5.4
1996	17.75
1997	26.05
1998 (estimate)	32.00
1999 (estimate)	38.00

Source: UN/ECE Working Party on Steel.

1.5 Evolution of "metallics" supply

Table 1.5 shows the evolution, during the last 25 years, of the various "metallics" or primary metals —scrap, pig iron in solid form as pig or in liquid hot metal and direct reduced iron or hot briquetted iron (DRI/HBI)— used to produce steel in both BOFs or EAFs.

Two remarks can be made about these data. Firstly, the figures for scrap consumption are not very precise and are probably too low as the data are often underestimated on the one hand, and on the other a number of countries are not included. Secondly, as regards pig iron, an important part is used in foundries, a point which is taken up in more detail in chapter 2.

The three primary metals have experienced different trends. Hot metal/pig iron production is not increasing (at least not in a significant way) but fluctuating; increasing in developing countries, especially in Asia, compensating, more or less, for the decrease in the older industrialized countries.

DRI/HBI production is recording a steady growth but is still at a relatively low level compared to pig iron and scrap. Trends in scrap are more complicated as they have to be related to the various scrap categories. These are analyzed in detail in chapter 2.

Table 1.5 Evolution in primary metals worldwide
(Millions of tonnes)

	Scrap consumption	Pig iron production	DR production
1966	...	335	...
7	...	352	...
8	...	376	...
9	...	412	...
1970	...	431	1
1	...	427	1
2	279	450	1
3	303	499	2
4	308	502	3
5	272	468	3
6	289	489	3
7	286	486	3
8	307	507	5
9	352	528	7
1980	333	508	7
1	332	498	8
2	300	453	7
3	313	458	8
4	318	461	9
5	350	475	11
6	331	468	13
7	363	480	14
8	372	510	15
9	377	516	16
1990	354	502	19
1	342	482	20
2	344	477	21
3	342	477	24
4	360E	483	27
5	365E	500 E	30
6	370E	490 E	33

Source: IISI *Steel Statistical Yearbook.*
E Estimate.
Note: The column "pig iron production" refers to pig iron for steel making.

Direct reduction (DR) is certainly the main "scrap substitute". Tables 1.6 and 1.7 give detailed information about its rapid expansion worldwide, i.e. DRI (direct reduced iron) and HBI (hot briquetted iron) compared to pig iron production. The distribution of DR production by processes indicates clearly the overwhelming domination of natural-gas-based processes which accounted for 92.3 per cent of total production in 1996 (table 1.8). Two processes —MIDREX and HYL (especially HYL III)— account, respectively, for 63.1 per cent and nearly 27.4 per cent of world production.

Developing countries and particularly those which have large resources of natural gas represent the most important share of DR producers (table 1.4 and figure 1.14). This shows clearly the importance of the cost of energy in the total cost of DRI/HBI.

If research and development in smelting reduction methods is progressing, for the time being only one process, COREX, has reached the commercial stage with 2 units in operation and 5 other in construction.

Table 1.6 Distribution of DR production worldwide (1996)

	Mt	Percentage
Industrialized countries	2.3	7.0
Western Europe		
United States and Canada		
Countries in transition		
Russian Federation	1.5	4.5
Developing countries		
Latin America	12.1	36.3
Middle East and North Africa	8.4	25.2
Asia	8.1	24.3
Africa	0.9	2.7
Sub-total	29.5	88.5
TOTAL	33.3	100.0

Sources: UN/ECE Working Party on Steel and IISI *Steel Statistical Yearbook*.

Table 1.7 Distribution of total primary iron production worldwide (1996)

	Mt	Percentage
Industrialized countries (Western Europe, United States, Canada and Japan)	238.4	45.7
Countries in transition (Eastern Europe and CIS)	79.2	15.2
Developing countries		
Latin America	32.0	6.2
Middle East and North Africa	10.0	1.9
Asia (excluding Japan)	161.5	31.0
Sub-total	203.5	39.1
TOTAL	521.1	100.0

Sources: UN/ECE Working Party on Steel and IISI *Steel Statistical Yearbook*.

Table 1.8 Distribution of DR production by process (1996)

	Mt	Percentage
by gas		
MIDREX	21.00	63.1
HyL I	2.81	8.4
HyL III	6.31	19.0
PUROFER	0.11	0.3
GHAEM	0.03	0.1
FIOR	0.33	1.0
Iron Carbide	0.12	0.4
Sub total	30.71	92.3
by coal		
Rotary kilns SLRN	1.08	3.2
Rotary kilns others	1.47	4.4
Kinglor Metor	0.02	0.1
Sub total	2.57	7.7
TOTAL	33.28	100

Source: MIDREX Corporation.

Table 1.9 Evolution of scrap resources and needs
(Millions of tonnes)

	1995	2000
Production of crude steel	750	822
Consumption of steel products	642	722
Consumption of primary metals		
Pig iron	515	551
Scrap	385	423
DRI	31	47
Scrap from the steel industry		
Circulating scrap	110	113
Capital scrap	11	14
Iron foundries	27	29
Total	147	156
Prompt scrap	82	87
Total	229	243
Residual scrap needs (obsolete and capital scrap)	157	180

Source: IISI, *Scrap and the Steel Industry*, Brussels, 1997.

Figure 1.13 World DRI production by process

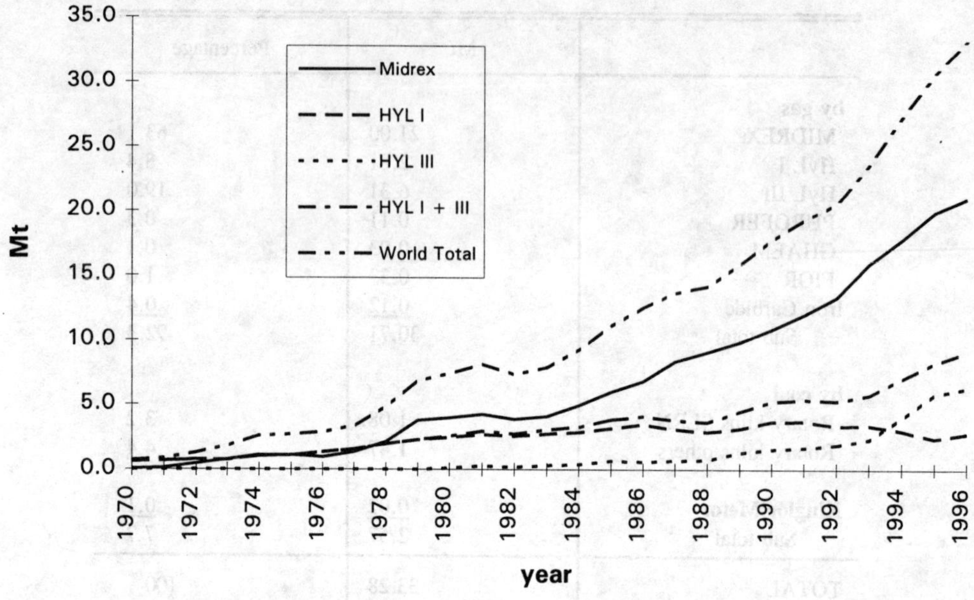

Source: UN/ECE Working Party on Steel.

Figure 1.14 World DRI production by region

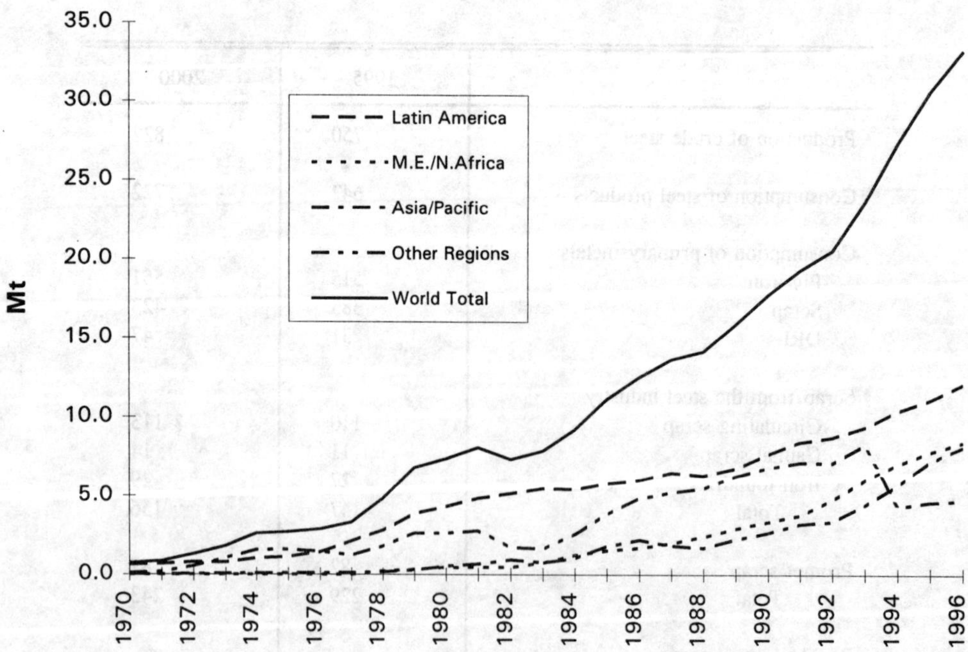

Source: UN/ECE Working Party on Steel.

1.6 The importance of the various scrap categories in the world steel industry

Finally, to summarize the role of scrap in the world steel industry, and also in iron foundry, figure 1.15 gives an overview of these industries for 1995, focusing on the importance of scrap. Such a picture tallies fairly well with similar estimates made by BHP (figure 1.16)

Figure 1.15 Overall scrap balance of steel industry worldwide

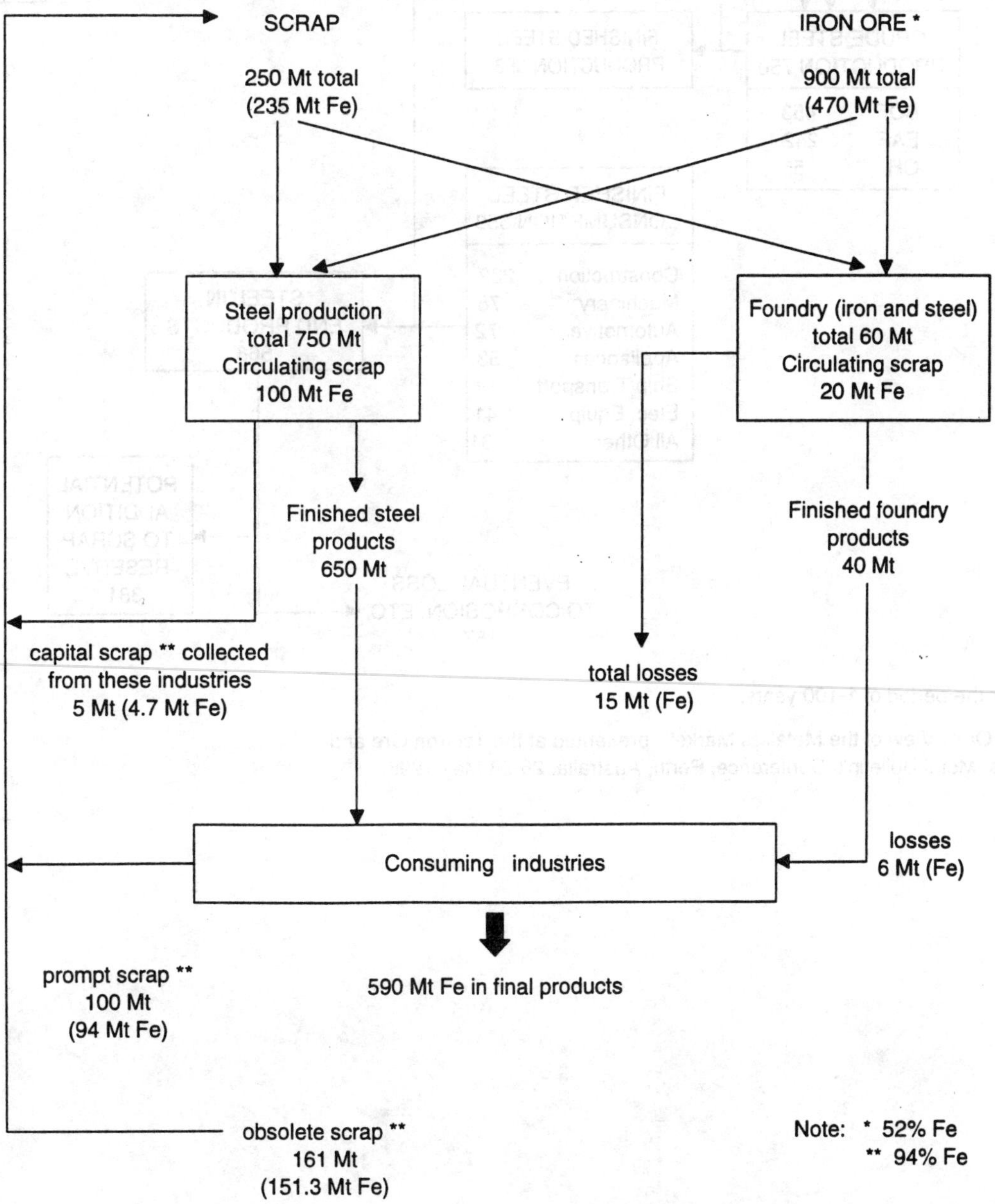

Figure 1.16 Simplified world ferrous material flow - 1995 estimate
(Millions of tonnes)

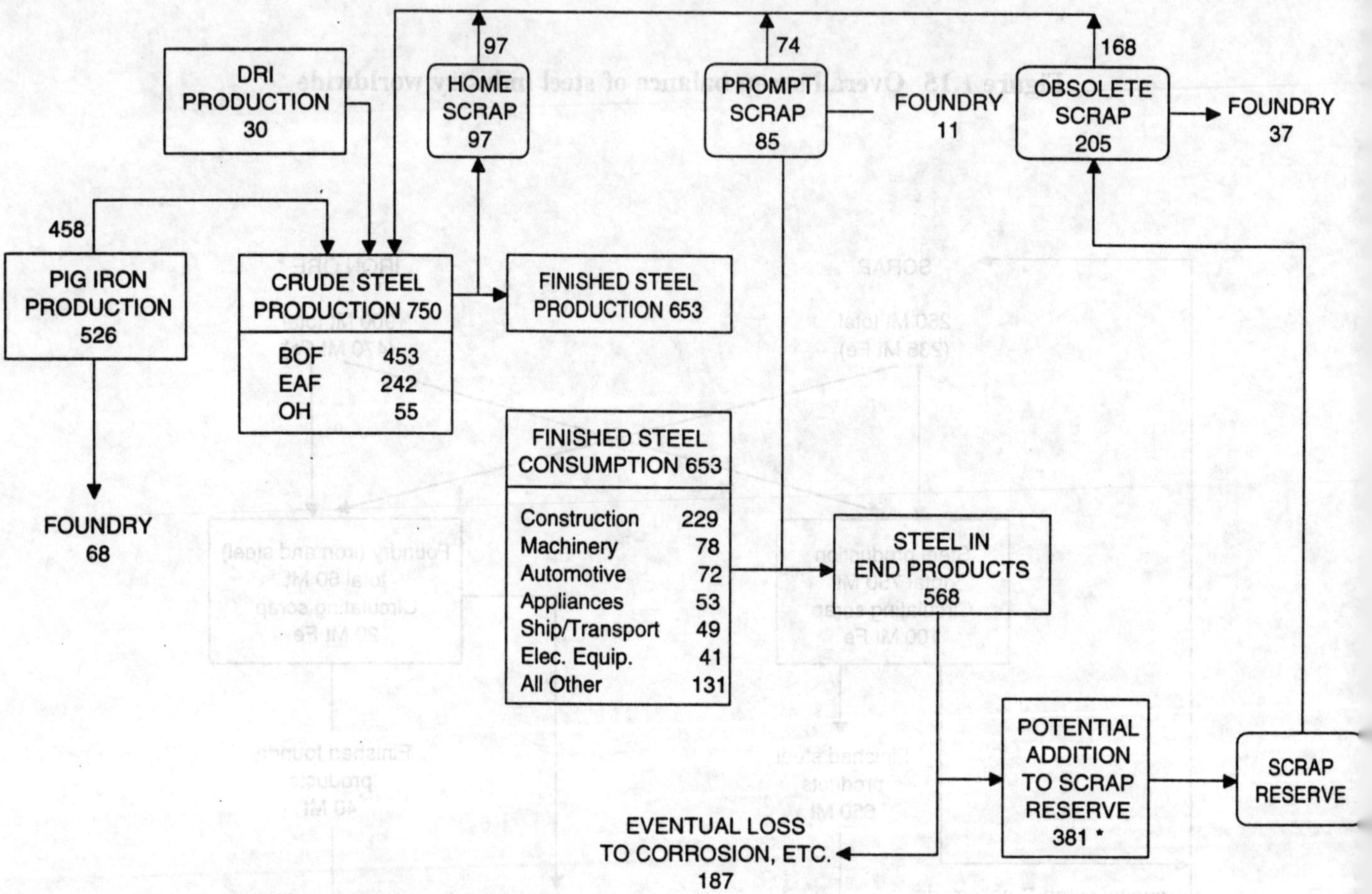

* Retrievable over the period of 1-100 years.

Source: "BHP Iron Ore's View of the Metallics Market", presented at the 1st Iron Ore and its Markets, Metal Bulletin's Conference, Perth, Australia, 26-28 May 1996.

CHAPTER 2

TRENDS IN SCRAP SUPPLY

The evolution of scrap supply, both by country and on the world scale, is analyzed under the five following headings:

- Own resources from the steel industry;
- Foundry resources;
- Capital scrap from the steel and foundry industries;
- Prompt scrap;
- Obsolete and capital scrap.

All categories and sources of scrap are indicated in figures 2.1 and 2.2.

2.1 Own resources from the steel industry

It is not necessary to detail again the correlation appeared in annex 2, covered in the last issue of this publication released, two years ago, between the evolution of the steel industry and, especially, the development of continuous casting and the decrease in circulating or own scrap. This trend is confirmed by the most recent data (see in annex, table A 2.2). The data from 15 countries of the EU (table 2.1) and a group of 11 other countries which answered the ECE questionnaire (table 2.2) show a similar important decrease of own resources from the steel industry. For the future, the predictions from all experts agree about the world trend of the first chapter (figures 1.15 and 1.16) which gave the following figures:

from about 200 Mt at the beginning of the 1970s[1]
to around 100 Mt in 1995[2]; and
towards 90 to 110 Mt in 2000 or 2005[3].

---

1 Jacques Astier in The Steel Industry in the Year 2005, IISI Brussels 1995.

2 Ibid. chapter 1 ; IISI, ITM/Battelle, BHP figures 1.15, 1.16 and 1.17.

3 Ibid.; IISI and ITM/Battelle.

Figure 2.1 Recycling and collection of own resources and prompt scrap
(Estimation for 1996)

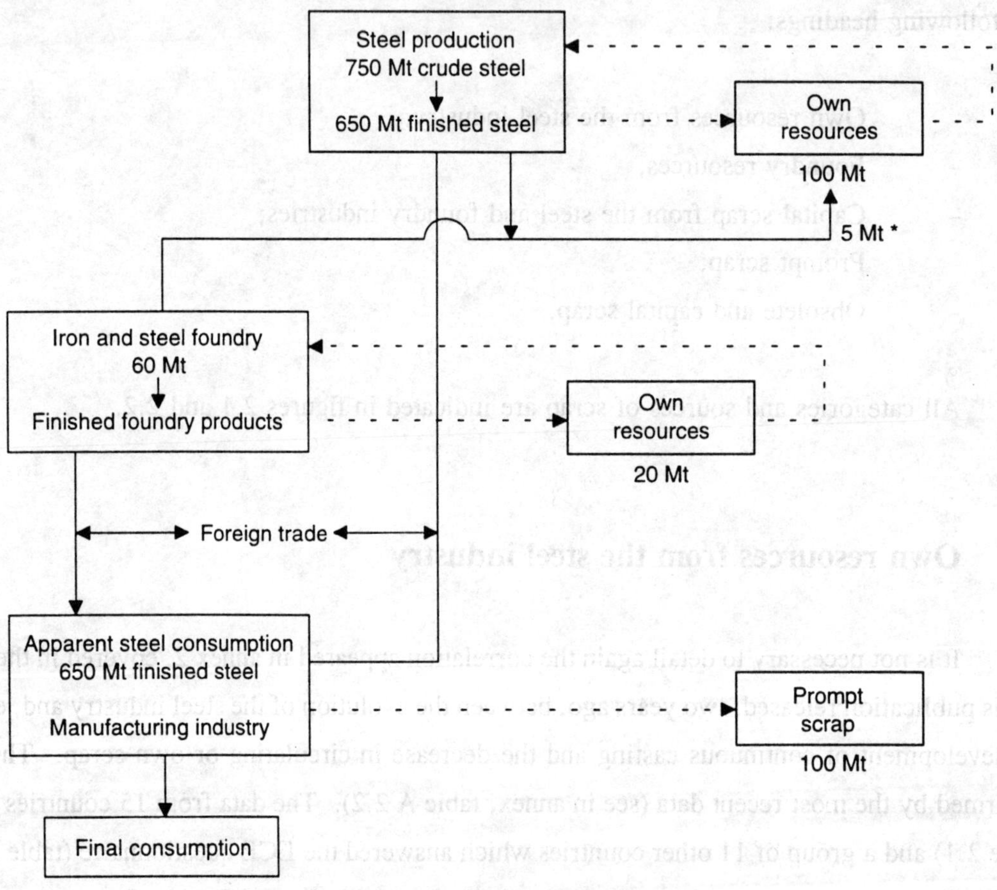

* 5 Mt of capital scraps are generated from demolition in steel works and foundries.

Source: UN/ECE Working Party on Steel.

Figure 2.2 Recycling and collection of obsolete and capital scrap
(Estimation for 1996)

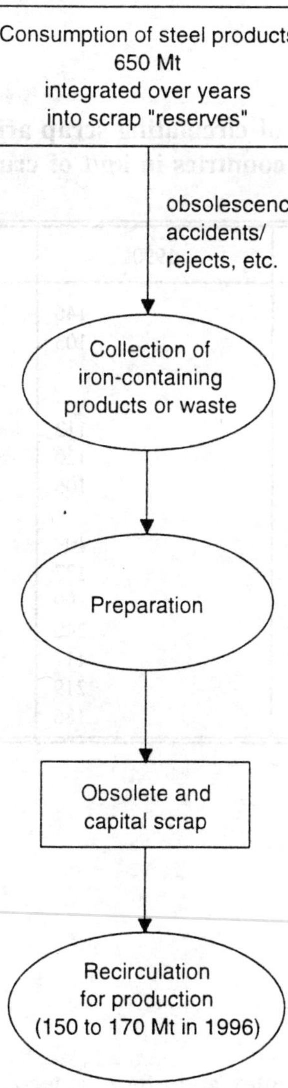

Source: UN/ECE Working Party on Steel.

For the whole world and in the future, especially given the trends outlined in chapter 1 regarding NNSC (Near Net Shape Casting), it is clear that the internal scrap resources of the steel industry, although this industry and its finished steel production will certainly develop and increase, are on a decreasing trend both in absolute quantity and in relative terms.

Table 2.1 Evolution of circulating scrap arisings in steelworks in EU 15 countries in kg/t of crude steel

	1985	1990	1995	1996
Austria	125	146	105	112
Belgium	149	103	100	102
Denmark	-	-	-	-
Finland	-	-	134	130
France	191	112	116	108
Germany	174	120	112	116
Greece	-	106	100	112
Ireland	-	-	29	38
Italy	140	107	81	83
Luxembourg	136	137	176	154
Netherlands	200	99	98	-
Portugal	164	240	340	141
Spain	-	117	-	-
Sweden	-	219	216	202
United Kingdom	-	186	137	148

Source: UN/ECE Working Party on Steel.

2.2 Foundry resources

This industry is somewhat complex and often dispersed but the trends are also clear as seen in the tables in the annex 1.

The flow sheet for the foundry, either for steel or for iron, is shown figure 2.3 and the main findings from the questionnaire as indicated in tables and of the annex can be summarized in tables 2.3, 2.4, 2.5, and 2.6.

Figure 2.3 Simplified flowsheet for the foundry industry

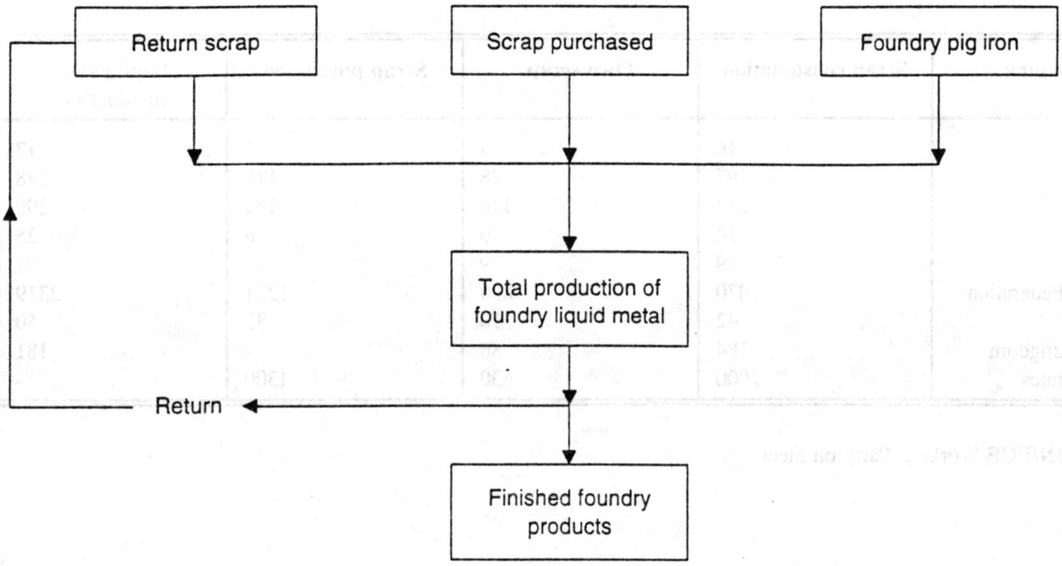

Source: UN/ECE Working Party on Steel.

Table 2.2 Evolution of circulating scrap arisings in steelworks in 11 countries in kg/t of crude steel

	1985	1990	1995	1996
Belarus	-	66	86	93
Bulgaria	-	55	45	42
Czech Republic	254*	304	233	215
Hungary	225	-	111	115
Japan	114	93	96	102
Korea, Rep. of	-	85	72	69
Poland	259	-	-	-
Romania	-	-	308	223
Russian Federation	226	307	220	223
Slovakia	-	153	112	103
United States	333	280	210	-

Source: UN/ECE Working Party on Steel.
* The former Czechoslovakia.

Table 2.3 Scrap balance for the steel foundry sector in selected countries (1996)

(Thousands of tonnes)

Country	Scrap consumption	Own scrap	Scrap purchased	Liquid steel production
Finland	16	5	11	17
France	197	78	119	228
Germany	284	126	162	299
Hungary	14	9	6	25
Italy	99	29	71	94
Russian Federation	2070	890	1231	2319
Slovakia	42	10	32	50
United Kingdom	184	86	-	181
United States	2000	630	1300	-

Source: UN/ECE Working Party on Steel.

Table 2.4 Scrap arisings in the steel foundry sector in selected countries

(Thousands of tonnes)

Country	1980	1985	1990	1995	1996
Finland	-	-	-	5	4
France	144	94	122	78	83
Germany	215	166	165	126	-
Hungary	-	-	20	9	-
Italy	71	-	83	29	-
Russian Federation	-	3500	1947	890	880
Slovakia	-	-	14	10	11
United Kingdom	-	-	113	86	90
United States	819	498	767	756	680
Total	4000E	4500E	3231	1989	2000E

Source: UN/ECE Working Party on Steel.
E Estimate.

Table 2.5 Scrap balance for the iron foundry sector in selected countries for 1995
(Thousands of tonnes)

Country	Scrap consumption	Pig iron consumption	Own scrap arising	Scrap purchased	Liquid Iron production
France	1899	750E	709	1190	2625
Germany	4211	700E	1590	2500	4748
Hungary	44	-	9	18	56
Japan	6438	1139	2617	3821	-
Korea, Rep. of	1418	605	328	971	-
Russian Federation	1783	1493	865	918	2316
Slovakia	29	19	13	16	59
United States	13000	1700	4901	8350	-

Source: UN/ECE Working Party on Steel.
E Estimate.

Table 2.6 Scrap arisings in the iron foundry sector in selected countries
(Thousands of tonnes)

Country	1980	1985	1990	1995	1996
France	835	653	715	709	683
Hungary	-	-	20	9	0
Japan	2171	2262	3001	2617	2625
Korea, Rep. of	-	-	307	328	343
Russian Federation	-	-	2976	865	860
Slovakia	550*	-	21	13	12
United Kingdom	848	-	693	-	-
United States	3904		5060	9961	-
Total scrap	13000E	14000E	12793	15200E	15200E

Source: UN/ECE Working Party on Steel.
* The former Czechoslovakia.
E Estimate.

It is clear that the role of scrap purchases by the foundry industry in overall scrap demand is limited, and is not growing, but is more important from the quality aspect as the specifications can be different and stricter than in the steel industry.

Regarding scrap arisings in the foundry industry, the values for the main countries which answered the ECE questionnaire on which report such activities can be estimated, excluding China which probably accounts for a further large amount at less than 2 Mt per year in the steel foundry industry, decreasing from probably around 4 to 5 Mt per year in 1980/1985, and around 15 Mt per year in the iron foundry industry, perhaps slightly increasing since 1980.

2.3 Capital scrap from the steel and foundry industries

Capital scrap from the iron and steel industry as well as from the steel and iron foundry industry is mainly collected during the demolition of obsolete facilities. Data from the answers to the ECE questionnaire are assembled in table A 2.4 and, in spite of the many variations from year to year and from country to country, it can be observed that the trend is toward a decrease in this resource, the amount is usually less than 10% of the total own resources of these industries, and probably with the present trend of building lighter and smaller production lines, a further decreasing trend in the future.

2.4 Prompt scrap

This is an important resource for the steel industry, including the foundry industry for the following three reasons:

(a) It is important in quantity with about 100 Mt/year;

(b) It is also important in terms of quality as it is usually low in non-ferrous metals ("residuals") but, perhaps more important, its chemical composition is known and is less varied than most obsolete scrap; and

(c) Its "production" or more precisely, its collection is continuous and linked with the evolution of steel production.

Unfortunately, little precise information is available from countries, mainly due to differences in the interpretation of the definition of circulating scrap from the steel industry: does it include, for example, cold rolling, coating, cold forming and finishing activities? The extent to which it covers various downstream activities, especially in the engineering and mechanical industries.

Table 2.7 Estimation of prompt scrap as a percentage of crude steel consumption in selected countries

Source	% of crude steel consumption
EC for:	
Germany	15.4%
France	14.8
Italy	12.8
Netherlands	12.5
Belgium - Luxembourg	12.2
United Kingdom, Ireland	15.0
Greece, Spain and Portugal	
United States	around 20%
Japan	11 to 13%
CIS (USSR) and Eastern European countries	13 to 16%
Developing countries	8 to 10%

Source: Government replies to the UN/ECE questionnaire.

2.5 Obsolete and capital scrap

This is the most important scrap resource, and will become more so in the future. However, it is difficult to give precise quantitative data as information has not been provided by many countries, and there is often confusion or, more exactly, different classification of prompt and obsolete or capital scrap in different countries.

To assess as fully as possible the level and evolution of total "collected scrap" (i.e. total prompt, obsolete and capital scrap), the total for each country will be taken from the line of DOMESTIC SCRAP MARKETED, which corresponds to this notion of total collected scrap in annex 1.

2.6 Total collected scrap by country

The data collected and assembled in the country total scrap balance (annex 1) can be analyzed in the following way: regarding western Europe and especially EU 15, table 2.8 gives relatively precise figures which, with estimates for the missing data, show the following trend:

41 Mt in 1980

42 Mt in 1985

47 Mt in 1990

56 Mt in 1995

57 Mt in 1996

Although a number of estimates are included, most of the data are precise and both the order of magnitude as well as the evolution are quite clear. It is, however, not so clear for various reasons in the case of the other countries of western Europe.

Regarding eastern Europe and CIS (see table 2.9), many figures are lacking but, from the information available, especially for eastern Europe, a total of 10 to 11 Mt can be estimated. For the CIS, estimate are more difficult and less precise. From the answers to the questionnaire, however, it would appear to be between 25 and 35 Mt per year.

For other regions (see table 2.10), the data are relatively precise for the United States, Japan, the Republic of Korea, Canada and Brazil but non-existent or imprecise for other countries or regions.

Estimates are made for the important east and south-east Asia region in table 2.11 and based on that estimates of world collected scrap are provided in table 2.12.

Tables 2.8 Total collected scrap in western Europe
(Thousands of tonnes)

Country	1980	1985	1990	1995	1996
Austria	731	658	802	1111	-
Belgium	1579	620	815	640	123
Denmark	-	-	-	-	-
Finland	-	-	410	343	552
France	7741	8224	9042	9600	9950
Germany	11712	12177	14349	20897	-
Greece	-	-	13	600E	-
Ireland	-	-	-	120	166
Italy	6156	7753	8197	9376	9198
Luxembourg	181	86	97	113	97
Netherlands	1548	1649	1445	1450E	-
Portugal	-	143	245	380	423
Spain	2993	-	3145	2200E	-
Sweden	1049	-	122	1243	1220
United Kingdom	7373	-	7902	8100E	-
Estimation of total	41063	42000E	47000E	56000E	57000E

Source: UN/ECE Working Party on Steel.
E Estimate.

Table 2.9 Total collected scrap in eastern Europe and the CIS countries
(Thousands of tonnes)

Country	1980	1985	1990	1995	1996
Azerbaijan	-	-	-	12	12
Belarus	-	-	969	900E	-
Bulgaria	-	-	-	585	552
Czech Republic	2400*	3017*	3000E	2500E	2500E
Slovakia	-	-	602	678	985
Hungary	-	1129	1150E	1268	1119
Poland	4786	3647	-	3900E	4000E
Romania	-	-	-	1751	2000E
Russian Federation	-	-	33675	19253	19193
Ukraine	-	-	-	3000E	3000E
Estimation of total	-	-	43000E	34000E	36000E

Source: UN/ECE Working Party on Steel.
* The former Czechoslovakia.
E Estimate.

Table 2.10 Total collected scrap in selected countries
(Thousands of tonnes)

Country	1980	1985	1990	1995	1996
Brazil	3305	3423	2765	2752	-
Canada	-	-	4700	5789	5880
Japan	27664	30961	36888	34692	36589
Korea, Rep. of	-	-	4495	9439	10568
Mexico	-	1542	-	-	-
United States	52233	43695	62907	67993	-

Source: UN/ECE Working Party on Steel.

Table 2.11 Total collected scrap in Asia
(Millions of tonnes)

Country	1980	1985
China	13	19
India	4.5	5
Indonesia	0.4	0.6
Korea, Rep. of	4.5	9.5
Malaysia	0.8	1
Pakistan	0.5	0.5
Philippines	0.5	0.6
Taiwan Province of China	3.2	5
Thailand	0.5	0.8
Total	27.9	42

Source: Euroconsult and Estimations by J. Astier.

Table 2.12 Estimate of total collected scrap worldwide in 1996
(Millions of tonnes)

Country	Collected scrap	Remarks
Western Europe (EU 15 + others)	61	see table 2.8
North America	75	see table 2.10
Japan	37	see table 2.10
Eastern and southern Asia	47	see table 2.11
Australia	2	-
Eastern Europe and CIS	36	see table 2.9
Latin America	6E	-
Middle East and Africa	6E	-
Estimation of total	270	-

Source: UN/ECE Working Party on Steel.

2.7 Conclusions

The evolution of the world scrap resources is summarized in table 2.13, referring to the estimates made two years ago and to a number of projections made for the future. What is clear is the decreasing trend in circulating scrap in the steel industry and probably in the foundry industry, as well as in the collection of capital scrap from industrial activities, and the increasing trend in the total of collected scrap (prompt, obsolete and capital scrap). What is still open to question is the exact trend in prompt scrap collection: it is certainly not increasing in the older industrialized countries but is, of course, increasing in the newly industrialized countries, especially in eastern and southern Asia. The main question concerns the evolution of obsolete and capital scrap collection: it is slowly increasing, depending on the country from the statistical data received in answer to the questionnaire. However, the exact amount of this increase in past years as well as in the future is subject to many questions such as the exact classification of the various scrap categories which leads, depending on the country, to different figures for own resources from the steel industry, prompt and obsolete scrap, and the exact amount of scrap for a large group of countries, especially the developing countries, which inevitably causes some uncertainty in the world estimates.

Table 2.13 Estimates of world scrap supply
(Millions of tonnes)

	Resources in 1994	Estimated resources in 1996	Forecasts for 2000	
			IISI estimates	ECE estimates
Circulating scrap in steel industry	110	100	113	100
Circulating scrap in iron and steel foundries	15	20	29	20
Capital scrap	-	5	14	5
Prompt scrap	85	100	87	120
Obsolete scrap	158	150-170	180	165
Estimation of total	368	375-395	423	410

Source: ECE, *Iron and Steel Scrap*, United Nations publication, Sales No.E.95.II.E.22, 1995.

CHAPTER 3

SCRAP DEMAND

The main scrap consuming processes can be split into two categories. The first category includes those which use limited and generally decreasing amounts of scrap: the blast furnace ironmaking process; the open-hearth steelmaking process; iron and steel foundries; and the oxygen steelmaking process. The second is the EAF (electric arc furnace) process which uses by far the largest amount of scrap and, furthermore, is increasing its scrap demand.

3.1 Blast furnace scrap demand

The data from the questionnaire are in annex tables A 2.12 and A 2.13. The result of this survey leads to three conclusions:

(a) The specific amount of scrap used in blast furnaces, in kilograms per tonne of pig iron, for the 16 countries which provided data (table A 2.13), varies from 0 to more than 100 kilograms. However, except in the special case of Hungary, the amount is generally only a few kilograms per tonne of pig iron;

(b) Although this point was not covered in the questionnaire, it is highly probable that the types and the quality of scrap used in BFs are such that they would be difficult to use in the steelmaking process;

(c) Although data are not available for all the countries, especially China and many developing countries, the total amount of scrap used in BFs worldwide is probably not more than 3 Mt, as was estimated in the UN/ECE study on iron and steel scrap for 1995.

3.2 Open-hearth furnace scrap demand

As explained in chapter one, the importance of this steelmaking process is decreasing and the answers to the UN/ECE questionnaire (as assembled in annex tables A 2.18 and A 2.19) include nine countries before 1990 and currently only five to which must be added the countries in table 3.1 to have a complete worldwide picture. The total production of around 50 Mt in 1996 will further decline but it is difficult to predict at what rate. UN/ECE predicts that it could be down to 40 Mt by the year 2000 but, due to the costs of replacing these steelmaking capacities with modern processes (oxygen or electric), the decrease in open-hearth steel production will probably be slower than was expected a few years ago.

Regarding scrap demand for this steelmaking process, the distinction has to be drawn between "Scrap process" operations, which are quite similar to conventional EAF operations and have a similar specific consumption of around 1,130 kilograms per tonne of steel, as reported in Latvia, and the more conventional "ore process" in which open-hearth operations use roughly equal amounts of scrap and pig iron with a small amount of iron ore. This appears to be the most common case for the remaining open hearths in operation in the world with an average scrap consumption of 500 to 600 kilograms per tonne of steel.

World demand, therefore can be estimated at 25 Mt of scrap in 1996 and a probable decrease to about 20 Mt in the year 2000 for open-hearth production.

3.3 Foundry industry scrap demand

As explained in chapter two, steel and iron foundries have to be considered separately.

As regards steel foundries, from IISI data on liquid steel for casting and answers to the questionnaire (table A 22 and data in table 2.2), it can be estimated from table 3.2 for western Europe and table 3.3 for the whole world that there is a total need of about 7 Mt of scrap per year.

Table 3.1 Production of steel in open hearth furnaces worldwide in 1996

(Thousands of tonnes)

Western and eastern European countries	
Czech Republic	60
Poland	1118
Romania	937
Turkey	719
Former U.S.S.R	
Azerbaijan	17
Georgia	44
Kazakhstan	526
Latvia	290
Russian Federation	17646
Ukraine	11750
Uzbekistan	130
Developing countries	
China	12500
Egypt	100
India	4300
Philippines	300
Total	50437

Source: UN/ECE Working Party on Steel, IISI Statistics.

As regards iron foundries, production is much higher but it is difficult to assess the scrap demand in countries or regions which have not provided data. The amount of scrap and pig iron (foundry iron) used to make one tonne of finished cast iron products differs widely from one plant or one country to another. Even if the basic flow sheet is the same (see figure 2.1), there may be many variations. The main data gathered from the ECE questionnaire are in table A.21 and A.22 in the annex. Despite the difficulties of assessing scrap demand for iron foundries, data and estimates for western Europe are given in table 3.4 and for the whole world in table 3.5.

It should be noted that the figure used for the total demand for scrap in iron foundries, the same one arrived at in the last ECE study on iron and steel scrap, is probably one of the less precise in this chapter.

Table 3.2 Steel foundry production and estimated scrap consumption in western Europe in 1996
(Thousands of tonnes)

Country	Production of Liquid Steel for Castings*	Scrap Consumption
Austria	7	-
Belgium	66	-
Finland	20	13
France	229	190
Germany	296	284[1]
Italy	107	99
Poland	32	-
Spain	123	-
United Kingdom	188	-
Total	1068	800E

Source: UN/ECE Working Party on Steel.
* From IISI Statistics.
E Estimate.

Table 3.3 Estimated production in steel foundries and scrap consumption worldwide in 1996
(Thousands of tonnes)

Country	Production of Liquid Steel for Castings*	Scrap Consumption
Western Europe	1100	800
Eastern Europe	400	300
CIS countries	3300	2000
North America	2000	2000
China	2200	1200
Japan	600	300
Others	300	200
Total	9900	6800

Source: UN/ECE Working Party on Steel.
* From IISI Statistics.

[1] Scrap consumption in 1995.

Table 3.4 Production of foundry iron and estimated scrap consumption in western Europe
(Thousands of tonnes)

Country	Production 1994	Scrap 1994	Consumption 1996
France	476	1827	1775
Germany	430	3815	4211 [2]
Italy	130	-	-
Subtotal	1036	5642	5985
Norway	70	-	-
Switzerland	110	-	-
Total	1216	6000 E	6000E

Source: UN/ECE Working Party on Steel.
E Estimate.

Table 3.5 Estimated production of foundry iron and scrap consumption worldwide
(Thousands of tonnes)

Country	Foundry Iron in 1994	Scrap Consumption	
		in 1994	in 1996
Western Europe	1216	6000	6000
Eastern Europe	543	1000	1000
CIS countries	1544	2000	2000
North America	444	13000 (1995)	-
China	24172	-	-
Japan	918	6330	6498
Others	1840	-	-
Total	30677	30000E to 40000E	30000E to 40000E

Source: IISI Steel Statistical Yearbook.

[2] Consumption in 1995.

3.4 Oxygen steelmaking scrap demand

To get a clear picture of scrap demand in BOFs (basic oxygen furnace) and of its future evolution, trends in oxygen steelmaking around the world have first to be considered. As shown in chapter one, it appears that oxygen steelmaking is still developing slowly in countries in transition, where it is replacing open-hearth steelmaking, and in a few developing countries like China. However, this increase is not compensating, at least for the time being, for the decrease in the older industrialized countries (North America, western Europe and Japan).

The second important factor is the amount of scrap used per tonne of steel. Table 3.6 gives an illustration of various practices that have been divided into three categories:

(a) The low-level ratio, about 100 kilograms of scrap per tonne of steel. This is typically the case in Japan for many years and now in the Republic of Korea, France and Italy.

(b) The medium-level ratio, of around 200 kilograms of scrap per tonne of steel.

(c) The high-level ratio, up to the practical metallurgical limit of the oxygen converter which is around 400 kilograms of scrap per tonne of steel. It should be noted that the country with the highest scrap ratio, Luxembourg, is logically abandoning this process and moving to 100 per cent EAF steelmaking.

The case of western Europe (table 3.7) indicates a slow decrease in the use of scrap in oxygen converters. The current estimates are more precise than those given in the last ECE scrap study and lead to about 19 Mt for the year 1994 for the whole of western Europe compared to an estimation of 18 Mt for 1993 two years ago. The total is estimated to decrease to 17 Mt in 1996 and probably 15.5 Mt in the year 2000.

For the whole world, the estimates shown in table 3.8 are for a total of around 80 Mt in 1996.

Table 3.6 Use of scrap in oxygen steelmaking
(Kilograms per tonne)

Country	1980	1985	1990	1995	1996
Low Level					
France	221	189	136	101	96
Japan	92	74	73	81	78
Korea, Rep. of	-	-	118	113	128
Italy	202	140	104	146	133
Medium Level					
Belgium	255	226	198	185	154
Germany	211	194	175	181	181
Netherlands	270	211	136	-	-
Portugal	-	172	222	212	194
Russian Federation	-	-	273	209	193
Spain	207	222	176	168	-
Sweden	165	151	151	193	200
Turkey	-	228	148	165	173
United Kingdom	260	-	174	-	-
High Level					
Austria	245	230	225	259	270
Hungary	-	-	263	264	274
Luxembourg	334	398	386	435	406
Romania	-	-	-	309	-
Slovakia	-	-	307	237	298
United States	323	296	277	282	274

Source: UN/ECE Working Party on Steel.

Table 3.7 Use of scrap in oxygen steelmaking by western Europe
(Thousands of tonnes)

Country	1990	1994	1996	Forecast for 2000
Austria	884	1051	1089	1000
Belgium	2057	1752	1420	1300
Finland	700E	700E	773	750
France	1856	1389	1052	1000
Germany	5499	5915	5321	5000
Italy	1165	1919	1388	1000
Luxembourg	1373	1022	474	0
Netherlands	707	932	900E	900
Portugal	41	87	87	100
Spain	986	981	850E	800
Sweden	409	555	637	650
Turkey	558	634	747	750
United Kingdom	2291	2203	2200E	2200
Total	18526	19140	16938	15450

Source: UN/ECE Working Party on Steel.
E Estimate.

Table 3.8 Total consumption of scrap in oxygen steelmaking worldwide
(Millions of tonnes)

Country	Oxygen Steel		Scrap Consumption in 1996
	1994	1996	
Western Europe	106	100.5	17
Eastern Europe	24	23	10
CIS countries	34	38.5	8
North America	68	69	18
Latin America	23	24	5
Africa and Middle East	9	8.5	1.5
Japan	67	66	5
Asia (without Japan)	84.5	112.5	15E
Oceania	8	8	1E
Total	423.5	450	80.5

Source: UN/ECE Working Party on Steel.
E Estimate.

3.5 Electric steelmaking scrap demand

Two main points influencing scrap demand in electric steelmaking have to be considered. Firstly, trends in steel production in EAFs worldwide (as indicated in chapter one), which show a rapid increase, and secondly the amount of scrap used in EAFs to produce one tonne of steel. This is examined in detail in tables A 2.16 and A 2.17 in the annex.

This second point is complicated by the fact that "scrap substitute" are beginning to play an important role. In the older industrialized countries where scrap is abundant, the conventional EAF practice is to operate with a charge made up exclusively of scrap which, based on the theoretical balance given in table 3.9, leads to a need of about 1,085 kilograms of scrap per tonne of steel. With high quality scrap with 97 per cent Fe content, the need decreases to 1,051 kilograms. It increases to 1,133 kg for low quality scrap with 90 per cent Fe content. This is more or less confirmed by the data in table 3.10 for western Europe and table 3.11 for eastern countries which answered the ECE questionnaire.

In many countries, the use of pig iron (sometimes in hot metal, or liquid state) is developing as indicated in tables 3.12 and 3.13.

In a number of developing countries where direct reduction is economical (see chapter 1, section 1.5), the burden of EAFs incorporates a large amount of DRI/HBI. In a recent survey[3], it was estimated that, in 1995 in the industrialized countries, the average EAF burden was 1.3% DRI/HBI, in the transition countries, it was 10.6% and, in the developing countries, it was 31.8%.

Summarizing the situation, tables 3.13, 3.14 and 3.15 give estimates for the year 1996. They clearly show the importance of "scrap substitutes" and the approximate worldwide demand by the EAF process.

Table 3.9 Theoretical estimate of "metallics" demand in an EAF steel production
(Kilograms per tonne)

Input	
1085 metallics of 94% Fe	1020
Ferroalloys addition	10
	1030
Output	1000
1000 tonnes of steel	20
Slag (0.085 of 25% Fe)	10
Dust (0.025 of 50% Fe)	1030

Source: UN/ECE Working Party on Steel.

---

[3] Jacques Astier; "The evolution of EAF feed worldwide", presentation at the Gorham/Intertech Conference, Charlotte, North Carolina, United States, 3-5 March 1997.

Table 3.10 Use of scrap in EAFs in western Europe

(Kilograms per tonne)

Country	1980	1985	1990	1995	1996
Austria	969	1011	735	1040	1012
Belgium	1056	1059	1019	989	962
Denmark	-	-	-	-	-
Finland	-	800	-	-	-
France	1010	1044	988	1047	1027
Germany	982	1031	927	975	1003
Greece	-	1095	1096	1140	1138
Ireland	-	-	-	1158	1142
Italy	1072	1090	1078	1026	1051
Luxembourg	-	-	-	1117	1152
Netherlands	-	-	-	-	-
Portugal	-	1117	1145	1165	1155
Spain	1093	1083	1095	1068	-
Sweden	1019	1130	991	984	994
United Kingdom	1052	-	1053	-	-
Croatia	-	-	1006	1073	1000
Norway	-	1061	1104	-	-
Slovenia	-	-	1148	923	865
Switzerland	-	-	1099	-	-
Turkey	-	-	1150	1150	1150

Source: UN/ECE Working Party on Steel.

Table 3.11 Use of scrap in EAFs in eastern Europe

(Kilograms per tonne)

Country	1980	1985	1990	1995	1996	1997
Bulgaria	-	-	1087	1132	1130	1111
Hungary	-	1094	1023	1145	1135	1059
Romania	-	-	-	790	-	-
Russian Federation	-	-	860	886	927	978
Slovakia	-	-	-	1409	1512	1250

Source: UN/ECE Working Party on Steel.

Table 3.12 Use of scrap in EAFs in selected countries
(Kilograms per tonne)

Country	1980	1985	1990	1995	1996	1997
Japan [1]	1045	1021	992	958	968	-
Korea, Rep. of [1]	-	-	1030	1065	1054	1051
Mexico [2]	416	413	260	288	305	174
United States	1048	1040	989	1013	929	-

Source: UN/ECE Working Party on Steel.

1 The decreasing consumption of scrap in Japan and the Republic of Korea is explained by the use of pig iron (see table 3.13).

2 Mexico's low consumption is explained by its use of DRI (see table 3.14).

Table 3.13 Use of pig iron in EAFs in selected countries
(Kilograms per tonne)

Country	1990	1995	1996	1997
Belgium	-	6	-	-
Bulgaria	49	29	32	31
Finland	46	43	41	59
France	2	40	-	-
Germany	58.	57	47	-
Greece	-	6	-	-
Hungary	41	11	33	-
Ireland	-	10	-	-
Italy	19	13	-	-
Japan	59	93	88	-
Korea, Rep. of	51	105	133	133
Romania	-	11	-	-
Russian Federation	33	55	60	62
Slovenia	85	2	14	-
Spain	1	-	-	-
Sweden	-	10	10	-
United States	-	31	40	-

Source: UN/ECE Working Party on Steel.

Table 3.14 Use of DRI in EAFs in selected countries
(Kilograms per tonne)

Country	1990	1995	1996	1997
France	0.2	-	-	-
Greece	-	0.2	-	-
Italy	3	2	-	-
Mexico	562	480	447	433E
Russian Federation	192	191	212	203E
Spain	6	26	-	-
United States	-	23	-	-

Source: UN/ECE Working Party on Steel.

E Estimate.

Table 3.15 Estimated world scrap requirement for EAF in 1996
(Millions of tonnes)

EAF steel production	245 Mt
Metallics requirements 1.085 (94% Fe) x 245 Mt	266
Estimate DRI/HBI	-30
Estimate pig iron	-10.5
Estimate world scrap requirement for EAF	226.5

Source: UN/ECE Working Party on Steel.

3.6 Conclusions

Total scrap demand is summarized in table 3.16. It can be observed from the table that there is clearly an increasing trend in world scrap demand, the demand is concentrated in order of importance, firstly on electric steelmaking, next, on oxygen steelmaking, and finally on the iron foundry industry, and the demand of the other sectors is either limited or decreasing.

Some imprecision is unavoidable in such a field where many data, for various reasons, have to be estimated. However, such figures show the general trends and give a good approximation of the reality.

Table 3.16 Estimated scrap demand worldwide
(Millions of tonnes)

	Demand in 1994	Demand in 1996	Forecast for 2000
Iron foundry	40E	40E	40E
Steel foundry	5	7	7
Blast furnace	3	3	3
Open hearth furnace	30	25	20
Oxygen steelmaking	76	80	75
Electric steelmaking	209	266*	300*
Total	363	421*	445*

Source: UN/ECE Working Party on Steel.
* Scrap substitutes are included.
E Estimate.

CHAPTER 4
SCRAP TRADE

Although practically all scrap with the exception of most circulating scrap is sold and purchased, the trade is mainly on a local or national basis and is covered, with the prices indicated in chapter 5, in national statistics. The present chapter deals with the international trade in scrap which can be split into: total volume of scrap trade; "local" scrap trade; regional scrap trade; and international maritime scrap trade.

4.1 Total volume of scrap trade

Most scrap collected is handled, prepared and sold by scrap dealers. The part which is recirculated is decreasing, as shown in chapter two and, consequently the total volume of scrap which is traded is expanding. If it is assumed that own resources are utilized on the spot and are not sold (which is largely true), scrap trade represents about 250 Mt per year out of a total consumption of 375 to 395 Mt of scrap per year (see table 2.13).

4.2 "Local" scrap trade

This category, local or more precisely "national" resources are shown in the national scrap balances by country (annex 1) under the line C. Domestic scrap marketed.

4.3 Regional scrap trade

Regional scrap trade refers to the internal trade of the following main regions of the world: North America, eastern Asia and Europe.

Regarding North America, mainly the United States, Canada and Mexico (to which could be added central America and the Caribbean) tables 4.1 and 4.2 indicate the matrix of imports and exports in the area. This trade has been relatively constant for years and represents exchanges along the northern and southern borders of the United States.

As regards the total exports of the United States, there is a slight decreasing trend but as can be seen in table 4.3, it is not yet clear if this is a fluctuation as in the past or a longer term trend. From the detailed data of the United States Geological Survey, export from the United States, outside of what may be considered regional (to Mexico and Canada, see tables 4.1 and 4.2) have shown the following trends:

Table 4.1 Regional scrap trade in North America in 1995
(Thousands of tonnes)

	Exports from				
To	United States	Canada	Mexico	Others	Total
United States	-	1610	165	315	2090
Canada	1650	-	-	-	-
Mexico	821	520	-	-	-
Others	8034*	-	85	-	-
Total	10505	2130	250	-	-

Sources: UN/ECE Working Party on Steel and United States Geological Survey.
* Details are further explained in table 4.3.

Table 4.2 Regional scrap trade in North America in 1996
(Thousands of tonnes)

	Exports from				
To	United States	Canada	Mexico	Others	Total
United States	-	1910	114	576	2600
Canada	1250	-	-	-	-
Mexico	1190	-	-	-	-
Others	6008	-	-	-	-
Total	8448				

Sources: UN/ECE Working Party on Steel and United States Geological Survey.
* Details are further explained in table 4.3.

Figure 4.1 Schematic map of regional scrap trade in North America in 1996
(Millions of tonnes)

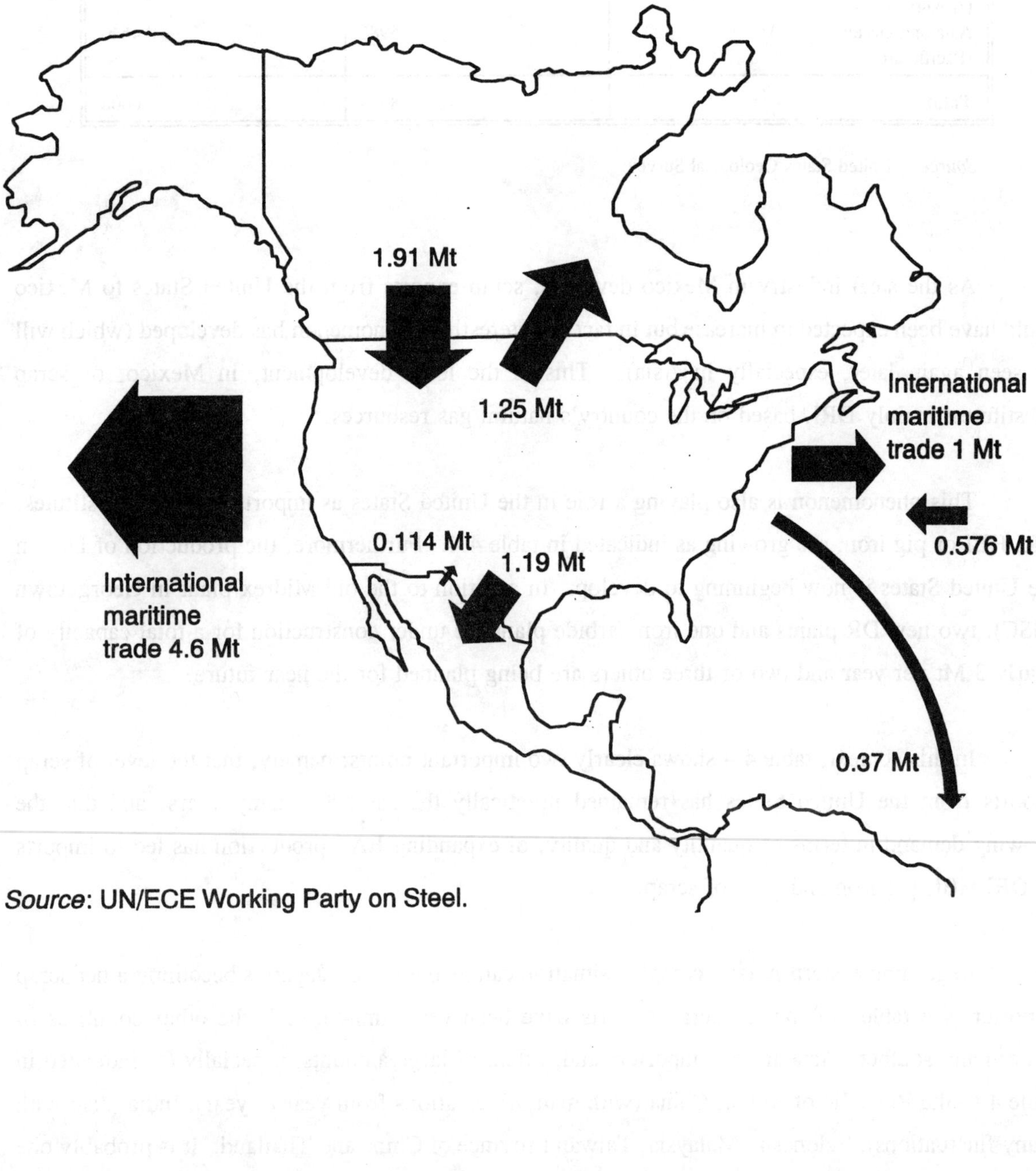

Source: UN/ECE Working Party on Steel.

Table 4.3 Scrap export of the United States to the rest of the world except North America
(Thousands of tonnes)

To	1995	1996
Venezuela	144	265
Other Latin America countries	168	106
Africa, Europe and Middle East (Atlantic area)	1736	997
Asia and Oceania (Pacific area)	5986	4640
Total	8034	6008

Source: United States Geological Survey.

As the steel industry in Mexico develops, scrap exports from the United States to Mexico could have been expected to increase but in fact an interesting phenomenon has developed (which will be seen again later, especially in Asia). This is the local development, in Mexico, of scrap substitutes, mainly DRI, based on the country's natural gas resources.

This phenomenon is also playing a role in the United States as imports of scrap substitutes, DRI/HBI or pig iron, are growing as indicated in table 4.4. Furthermore, the production of DRI in the United States is now beginning to develop. In addition to the old Midrex plant in Georgetown (GSC), two new DR plants and one iron carbide plant are under construction for a total capacity of nearly 3 Mt per year and two or three others are being planned for the near future.

In this context, table 4.4 shows clearly two important points; namely, that the level of scrap exports from the United States has remained practically the same for many years, and that the growing demand in terms of quantity and quality, of expanding EAF production has led to imports of DRI/HBI, pig iron and even of scrap.

Regarding eastern Asia, a complex situation can be observed. Japan is becoming a net scrap exporter (see table 4.5) but its scrap exports have been very limited. All the other countries of eastern and southern Asia are net importers and, often, of large amounts, especially (as indicated in table 4.6) the Republic of Korea, China (with many fluctuations from year to year), India (also with many fluctuations), Indonesia, Malaysia, Taiwan Province of China and Thailand. It is probably one of the regions for which precise data is not yet available. The only exporters of small volumes are Hong Kong and Singapore.

It should be emphasized again that for this region many data are missing, as only a few countries responded to the UN/ECE questionnaire. And since consumption and production of steel are developing very fast, it is even more difficult to arrive at precise figures. However, the picture is beginning to be clearer with an important expansion of scrap imports as indicated in table 4.6 but, as was mentioned for Mexico, Latin America and the United States, a compensating phenomenon is taking place and imports seem to have decreased since 1993. This is clearly related to the local development of the "scrap substitutes", through the building of both DR plants and blast furnaces, especially mini blast furnace.

Finally, in Europe, we also have a complex situation. Although western Europe is more or less clear, the countries of eastern Europe raise many questions.

Table 4.4 Increase in imports of primary metals in the United States
(Thousands of tonnes)

Year	Scrap exports	Scrap imports	Imports			
			Pig iron		DRI	
			Total	Net*	Total	Net*
1980	10131	506	-	-	-	-
1985	9026	545	-	-	-	-
1990	11580	1285	-	-	-	-
1993	9810	1390	828	801	1090	1073
1994	8810	1740	2500	2444	1170	1153
1995	10400	2090	2360	2306	1190	1185
1996	8445	2710	2656	2596	1181	1121

Sources: UN/ECE Working Party on Steel and United States Geological Survey
* Indicates imports-exports.

Table 4.5 Trends in scrap trade in Japan
(Thousands of tonnes)

	1980*	1985*	1990	1995	1996
Exports	159	166	396	915	1993
Imports	2896	3254	1048	1209	375
Net total (exports-imports)	-2737	-3088	-652	-294	1618

Source: UN/ECE Working Party on Steel.

Table 4.6 Scrap trade in eastern and south-eastern Asia
(Thousands of tonnes)

Imports - Exports +	1985	1990	1993	1994	1995	1996
China	-2012	-47	-3130	-2203	-1236	-
Hong Kong	+299	+398	+311	+200	+235	-
India	-1795	-3022	-2573	-1089	-1641	-
Indonesia	-160	-1177	-1873	-1011	-1462	-
Korea, Rep. of	-3478	-3845	-4893	-4687	-5006	-6490
Malaysia	-12	-684	-1417	-869	-1064	-
Pakistan	-114	-456	-354	-185	-121	-
Philippines	-1	-65	-53	-28	-53	-
Singapore	+129	+79	+24	+124	+101	-
Taiwan Province of China	-270	-1218	-1736	-1280	-673	-
Thailand	-654	-1091	-1458	-1124	-1035	-
Total (without Japan)	-8068	-11128	-17152	-12152	-11955	-

Source: UN/ECE Working Party on Steel.

Regarding western Europe, the detailed tables 4.7, 4.8 and 4.9 indicate that, using the definition of IISI for the countries for this region, there is a constant scrap deficit. The EU has been quite close to an equilibrium but with its expansion from 6 to the current 15 countries, there are increasing scrap needs. Moreover, the "other western European countries" and particularly Turkey are adding to growing scrap needs. Another problem is that, depending on the statistical sources, there are differences in the order of several million tonnes. Finally, the influence of the external trade of western Europe cannot be ignored, especially the imports from eastern Europe and exports to Asia.

As regards eastern Europe and the CIS, the number and quality of data are improving, especially for eastern Europe, but are still somewhat questionable for the CIS as indicated in table 4.10. In fact, it is highly probable that scrap exports from eastern Europe and the CIS are underestimated.

Finally, the map in figure 4.2 shows in a schematic way, with some simplifications and some uncertainties, the complex situation regarding scrap trade in Europe.

Table 4.7 Imports of scrap in western Europe

(Thousands of tonnes)

Country	1990	1995	1996
Austria	55	529	500E
Belgium	1855	3745	3811
Denmark	74	169	-
Finland	46	297	223
France	980	1627	1273
Germany	1041	1017	1197
Greece	1752	335	1750E
Ireland	217	178	213
Italy	5807	6337	4524
Luxembourg	629	1402	1554
Netherlands	1160	3050	1200E
Portugal	83	60	35
Spain	4291	4883	5000E
Sweden	222	282	269
United Kingdom	26	176	247
Sub-total for EU 15	18238	24087	21796E
Croatia	33	2	2
Norway	6	231	216
Slovenia	-	260	-
Sweden	136	90	-
Turkey	3650	6852	7000E
Former Yugoslavia	411	51	-
Total	22474	31573	29014

Source: UN/ECE Working Party on Steel.
E Estimates.

Table 4.8 Exports of scrap in western Europe
(Thousands of tonnes)

Country	1990	1995	1996
Austria	343	489	-
Belgium	809	1976	1741
Denmark	310	437	-
Finland	2	180	228
France	3470	3680	3740
Germany	5043	7965	8000E
Greece	46	39	-
Ireland	26	33	-
Italy	19	33	18
Luxembourg	-	24	33
Netherlands	2594	3622	2500E
Portugal	15	91	58
Spain	23	19	-
Sweden	54	263	265
United Kingdom	2807	3442	3419
Sub-total for EU 15	15561	22293	20000E
Croatia	-	100	106
Iceland	-4	18	-
Malta	-9	10	145
Norway	12	97	
Slovenia	-	48	-
Sweden	66	312	-
Turkey	7	13	16
Former Yugoslavia	218	10	-
Total	15851	22901	20267E

Source: UN/ECE Working Party on Steel.
E Estimate.

Table 4.9 Net scrap imports in western Europe
(Thousands of tonnes)

	IISI	UNCTAD
1975	3968	
1980	5365	
1985	4910	4910
1986	3788	
1987	4521	
1988	3921	
1989	4451	6670
1990	4226	6575
1991	3207	
1992	3397	
1993	2622	
1994	5861	
1995	-	10670

Table 4.10 Net scrap imports (-) and exports (+) in eastern Europe
(Thousands of tonnes)

Country	1980	1990	1995	1996
Bulgaria	-	+1	+11	-
Czech Republic	-	-	+700	+821
Hungary	-	-411	+748	+587
Poland	-153	+423	+448	+438
Romania	-	-	+8	+27
Slovakia	-	-	-38	+54
Sub-total	-153	+13	+1877	+1927
Estonia	-	-	+251	+186
Latvia	-	-	+1	+6
Lithuania	-	-	+504	+579
Sub-total	-	-	+756	+771
CIS countries	-	-	+2000E	+3000E
Total	-	-	+4633E	+5698E

Source: UN/ECE Working Party on Steel.
E Estimate.

To complete the picture, tables 4.11 and 4.12 show the situation in two regions which have some importance for scrap trade. Latin America including Mexico which is, on the whole, a small scrap importer accounting for about 1 Mt/y, and Africa, which is a negligible scrap trader, with exports of nearly 0.2 Mt/y.

Table 4.11 Evolution of scrap trade in Latin America - Net trade (+ exports, - imports)
(Thousands of tonnes)

Country	1995	1990	1994	1995*
Argentina	-1	-21	-2	+4
Brazil	-104	-113	-148	-147
Colombia	-58	-104	-28	-218
Peru	-	-	-22	-2
Venezuela	-496	0	-2	-80
All others	-	-	-	+41
Sub-total	-659	-238	-202	-402*
Mexico	-	-712	-617	-625*
Total	-	-950	-819	-1031*
		-962*	-1026*	-1031*

Source: IISI, except for (*) which were taken from UNCTAD, *Handbook of World Materials Trade Statistics*, 1990-1995.

Figure 4.2 Schematic map of regional scrap trade in Europe in 1995
(Millions of tonnes)

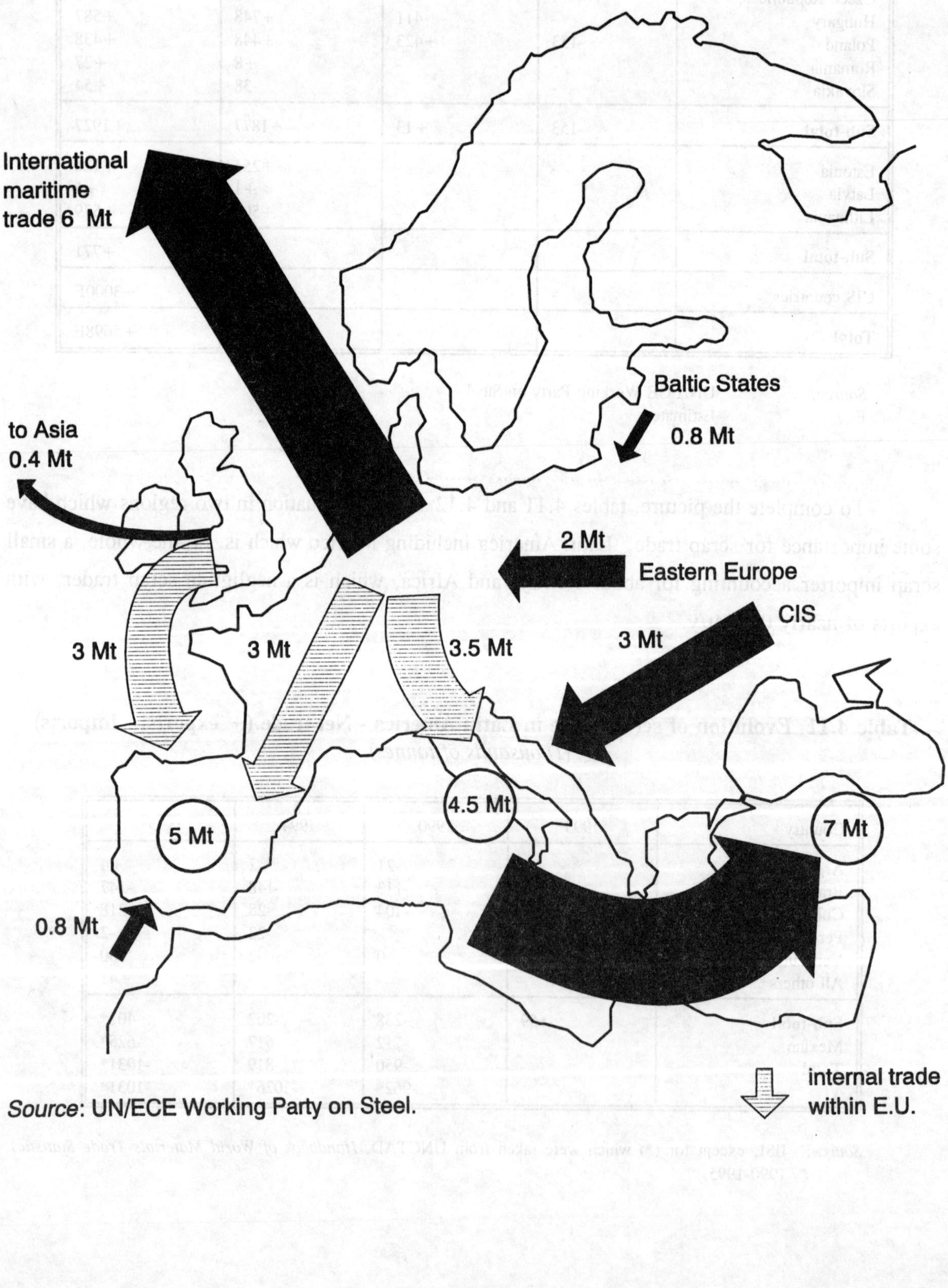

Source: UN/ECE Working Party on Steel.

Table 4.12 Evolution of scrap trade in Africa - Net trade (+ exports, - imports)
(Thousands of tonnes)

Country	1995	1990	1994	1995*
Egypt	-	-	-70	-
South Africa	+1	-	+78	-
Other	-	-	-	-
Total	+1	- -47*	+8 +160*	- +192*

Source: IISI, except for (*) which were taken from UNCTAD, *Handbook of World Materials Trade Statistics*, 1990-1995.

4.4 International maritime scrap trade

Before attempting to make an assessment, it is necessary to comment on the problems of scrap statistics. IISI data indicate an excess of 8 Mt/y of world scrap imports over world scrap exports. UNCTAD, with slightly different data, shows a similar discrepancy (see table 4.15).

Such difficulties do not prevent a general picture of world maritime scrap trade being formed (see figure 4.3) but the lack of explanation for the 5 Mt/y missing in the overall balance has to be pointed out (see table 4.13):

Suffice it to say that world maritime trade, even if Turkey with 7 Mt is included, is not developing significantly and is of the order of 20 Mt/y, more or less the same level as a few years ago.

Table 4.13 International maritime trade in main regions in 1996
(Millions of tonnes)

Main scrap exporting regions	per year	Main scrap importing regions	per year
North America	9.0	EU 15	0.8
Eastern Europe	2.6	Other western Europe	0
CIS	2.6	Turkey	6.8
Australia	0.35	Latin America	1.0
South Africa	0.3	Eastern and Southern Asia	11.5
Middle East	0.3		
Other Africa	0.2		
Japan	0.2		
Total	**15.55**		**20.1**

Source: UN/ECE Working Party on Steel.

Figure 4.3 Schematic map of international maritime scrap trade in 1995
(Millions of tonnes)

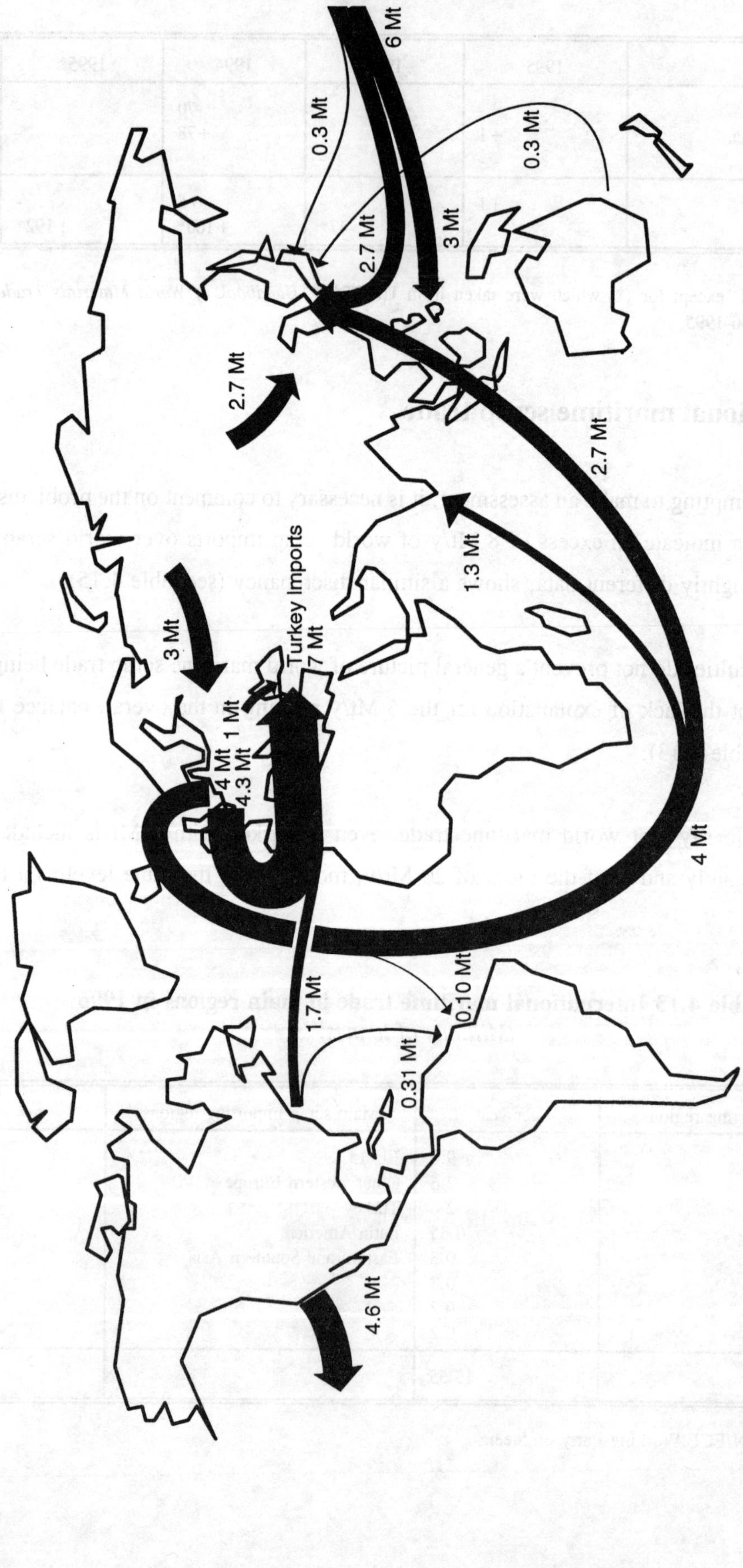

Source: UN/ECE Working Party on Steel.

Table 4.14 World scrap trade in 1995
(Thousands of tonnes)

	Exports	Imports	Net
Developed market economies	36879	28407	+8472
Canada	2129	1633	+496
United States	10439	2060	+8379
EU 15	22276	23154	-878
Other Europe	427	321	+106
South Africa	330	10	+320
Japan	915	1209	-294
Oceania	363	20	+343
Developing countries and territories	2839	21830	-18991
Latin America	519	1550	-1031
Africa	293	100	+193
Asia	1858	19869	-18011
Former Yugoslavia	169	311	-142
Eastern Europe	5702	460	+5242
World	45420	50697	-5277

Source: UNCTAD, *Handbook of World Materials Trade Statistics*, 1990-1995.

Table 4.15 World scrap trade in 1994
(Thousands of tonnes)

	Exports		**Imports**	
	UNCTAD	**IISI**	**UNCTAD**	**IISI**
Western Europe	23149	22056	28066	27917
USA + Canada	10474	10474	3549	3603
Japan	967	969	1068	1069
Oceania	442	441	33	63
Sub-total	35032	33940	32716	32652
Latin America	455	132	1481	951
Africa	268	177	107	169
Asia	1810	177	20251	11584
Other developing countries	3	-	329	-
Sub-total	2536	486	22168	12704
Eastern Europe	2378	1391	298	372
+ CIS countries	2335	1653	277	260
+ Baltic States	708	-	-	-
Sub-total	5421	3044	575	632
Total	42989	37470	55459	45988

Source: IISI and UNCTAD.

CHAPTER 5

TRENDS IN THE PRICE OF SCRAP

The subject of scrap price and especially its variations both in various countries and regions of the world and over the years or even shorter periods of time is very important and is perhaps the most important subject of this study as it is closely related to the various other subjects of demand, supply and trade. It is covered here under the following six headings:

- Evolution of scrap prices in several important regions;
- Influence of exchange rates;
- Interpretation of short- and medium-term scrap price fluctuations;
- Local and international trade;
- The different scrap categories;
- General comments about the evolution of scrap prices.

5.1 Evolution of scrap prices

The evolution of scrap prices is shown in figures 5.1, 5.2 and 5.3, for the three countries which are the most important scrap collectors and users. In figure 5.1, the United States for Heavy Metal Composite Scrap Prices, in figure 5.2, Germany — quite representative of Europe — for German New Steel Scrap Sorten No 2 Stahlneuschrott), and in figure 5.3, Japan for Japanese Heavy Steel Prices (H 2).

It is examined how prices were affected by exchange rates, and short- and medium-term fluctuations, and what differences there were in the various categories of scrap and in local versus international trade.

Figure 5.1 Trends in the United States No.1 heavy metal composite scrap price
(US$/gross ton)

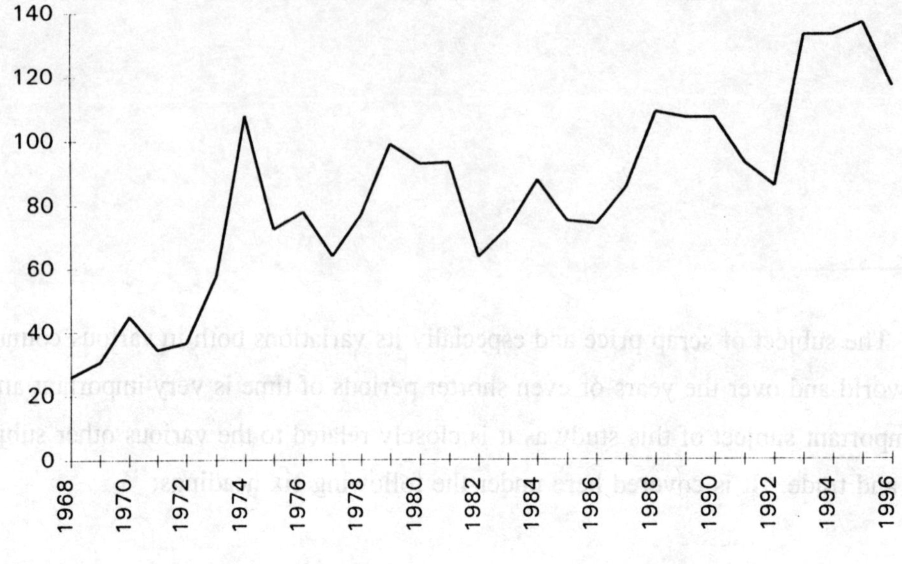

Source: American Metal Market.

Figure 5.2 Trends in Germany's new steel scrap price (No. 2, December)
(DM/metric tonne)

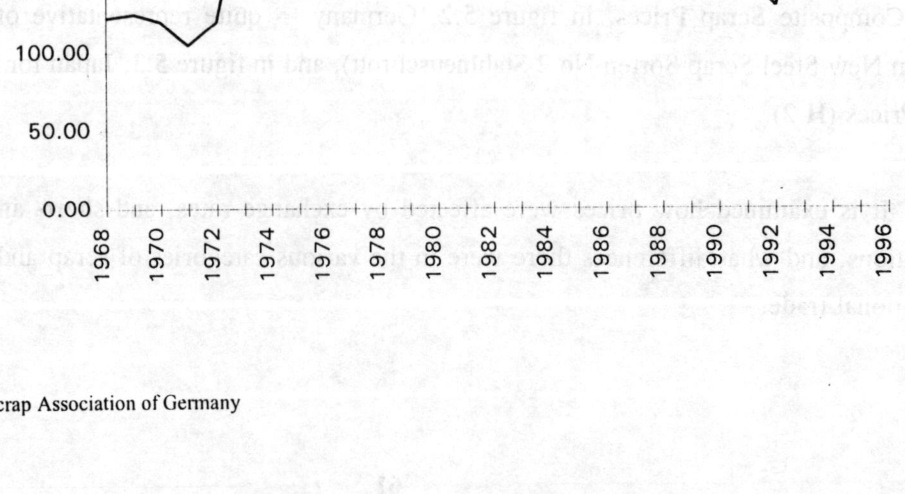

Source: Scrap Association of Germany

Figure 5.3 Trends in Japan's heavy steel scrap price (H-2, December)
(Yen/metric tonne)

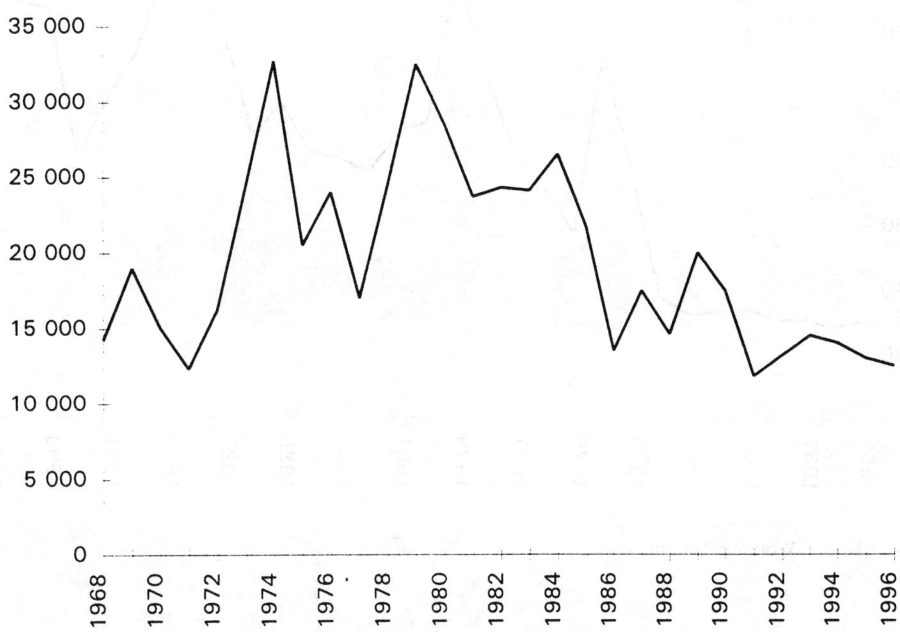

Source: Japan Iron and Steel Federation.

5.2 Influence of exchange rates

The prices indicated in Deutch Marks and Yen, in figures 5.2 and 5.3 have been converted into American dollars using the corresponding exchange rate for each year (figure 5.4 and 5.5). This confirms that, when expressed in a single currency, i.e. US$ (figure 5.6), the evolutions are quite similar and indicates that the scrap market, at least in the main regions collecting and using scrap, is a single entity. The influence of transportation (see 5.4) and various categories of scrap are then considered (see 5.5).

Figure 5.4 Trends in Germany's new steel scrap price (No. 2, December)
(US$/metric tonne)

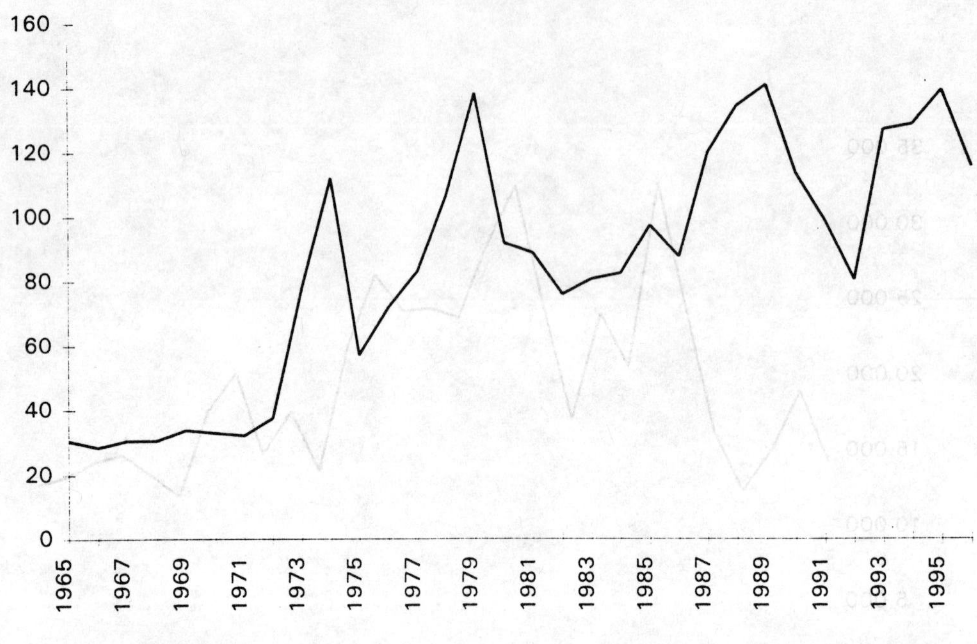

Source: UN/ECE Working Party on Steel.

Figure 5.5 Trends in Japan's heavy steel scrap price (H-2, December)
(US$/metric tonne)

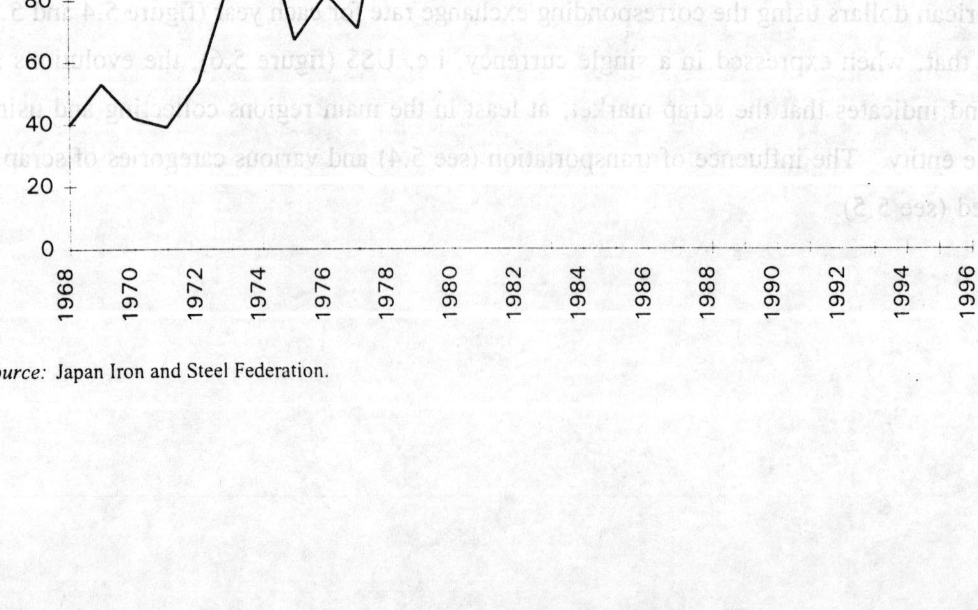

Source: Japan Iron and Steel Federation.

Figure 5.6 Comparison of price of scrap in the United States, Germany and Japan
(US$/metric tonne)

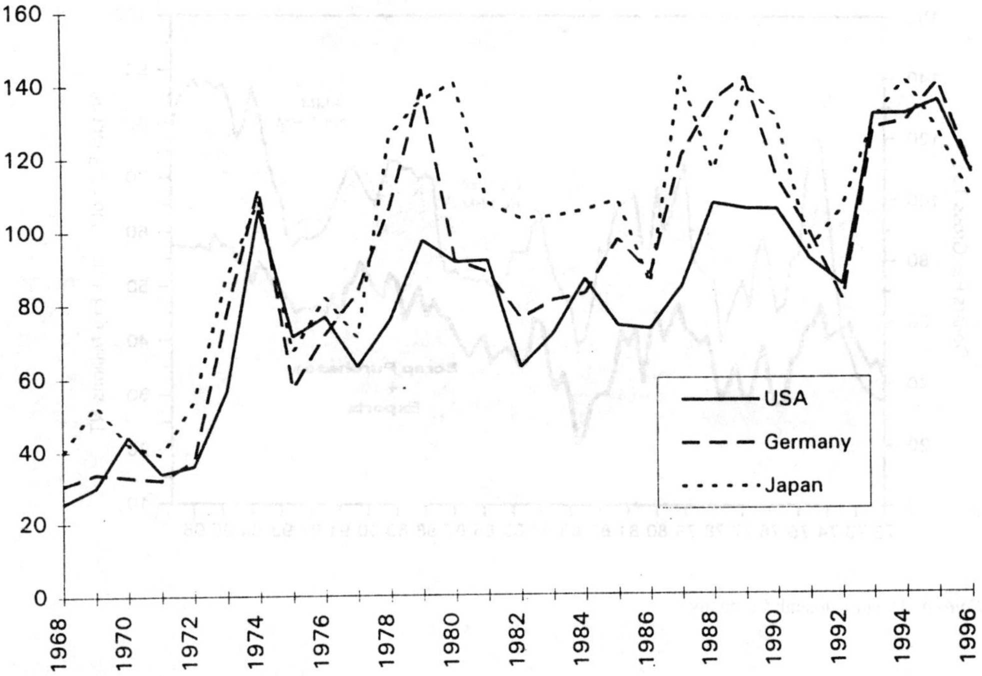

Source: UN/ECE Working Party on Steel.

5.3 Interpretation of short- and medium-term scrap price fluctuations

In theory, as well as in practice, with quotations every month in Europe and, every week in the United States, trade in scrap, as indicated in the figures 5.1 to 5.3 for example, can be considered a "true market" where a balance is reached, with each quotation between the demand from the users i.e. the steel industry and the supply from the scrap traders.

But, it is more complicated as there are various types of steel users in the steel industry as seen in chapter 3 and, on the other hand, the collection and trade in scrap is complex. Recent analysis in the United States[1] confirms that the economic situation and, as an indicator, the demand for steel products (in the United States market in this case, plus, of course, net scrap trade) has a clear relation with the evolution of scrap prices (see figures 5.7, 5.8, 5.9, and 5.10). This is the main factor explaining price fluctuations.

1 Gorham/Intech Conference, Charlotte, North Carolina, United States, 3-5 March 1997, especially the contribution by Wulff and Morrison.

Figure 5.7 United States steel mill scrap purchases and exports versus scrap prices
(US$/gross tonne)

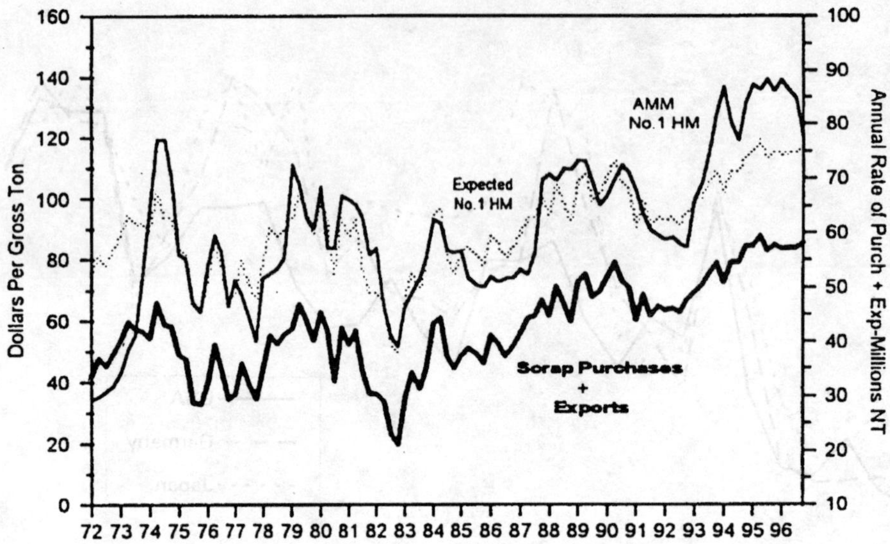

Source: David J. Joseph Company.

Figure 5.8 Total United States scrap demand and scrap costs (1980-1995)
(Millions of net tonnes)

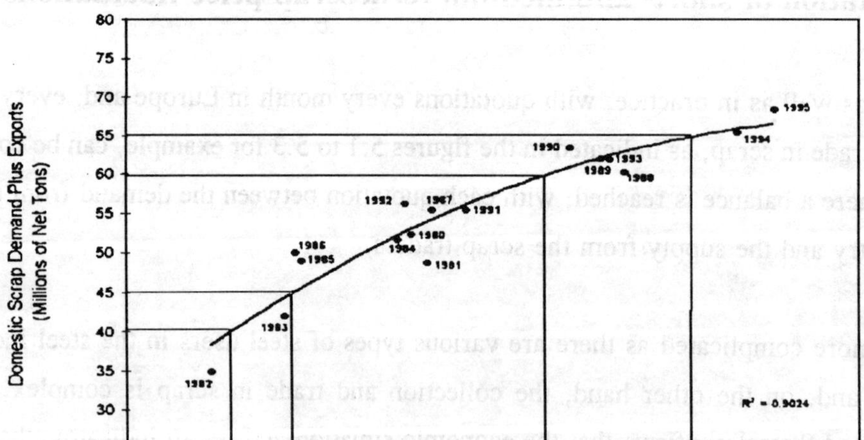

Source: David J. Joseph Company.

Figure 5.9 Scrap consumption versus steel production
(Millions of metric tonnes)

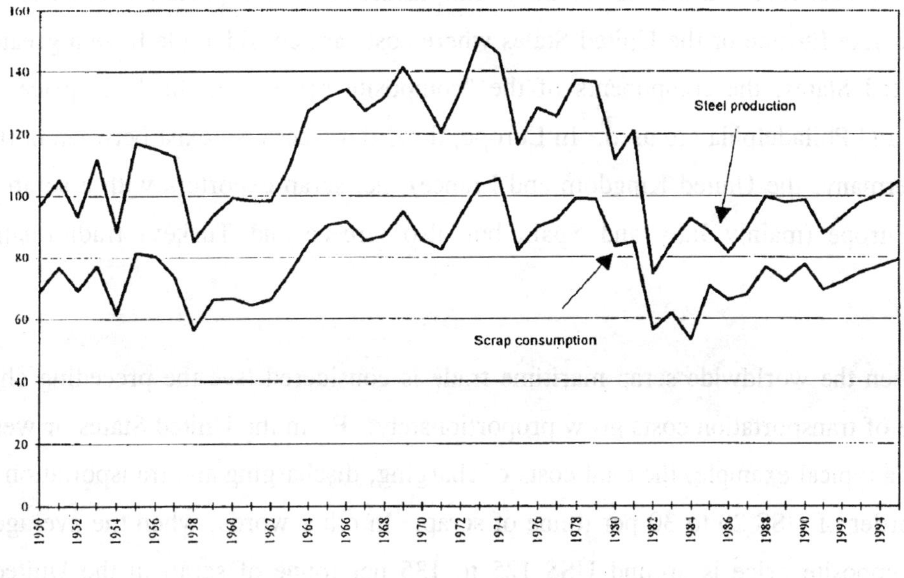

Source: David J. Joseph Company.

Figure 5.10 Scrap steel price (1960-1997)
(US$/gross tonne)

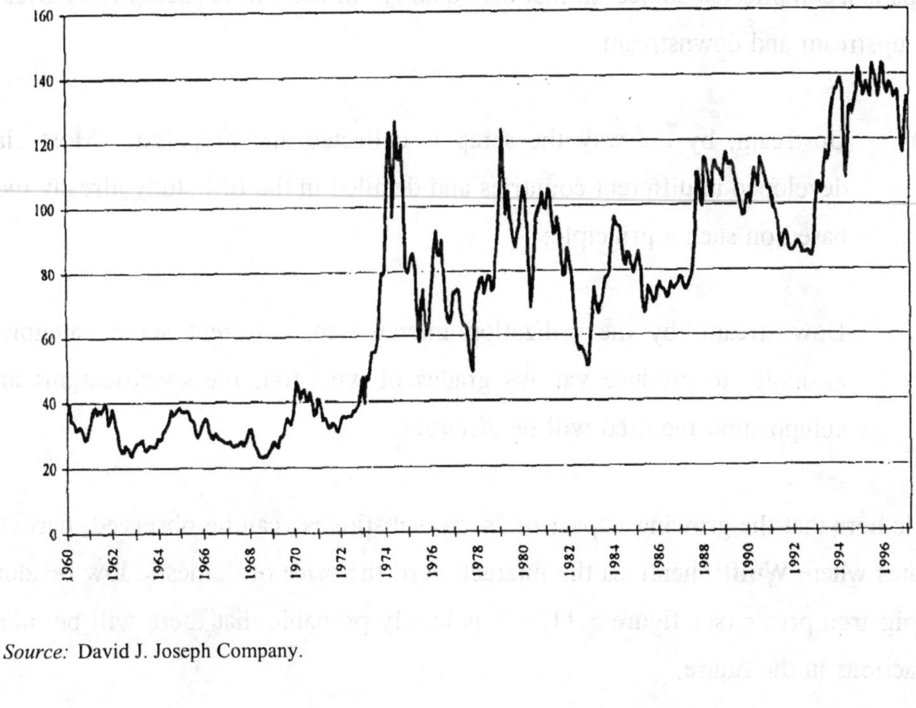

Source: David J. Joseph Company.

5.4 Local and international trade

On a local level, in a small country the influence of transportation costs is ignorable but larger areas like Europe or the United States where costs are considerable have a greater influence. In the United States, the components of the "composite price" i.e. the local prices in Chicago, Pittsburgh and Philadelphia are used. In Europe, the differences also exist between northern Europe (mainly Germany, the United Kingdom and France), net scrap exporters with a scrap surplus, and southern Europe (mainly Italy and Spain but also Greece and Turkey) traditionally net scrap importers.

When the worldwide scrap maritime trade is considered (see the preceding chapter 4), the importance of transportation costs grow proportionately. From the United States or western Europe to Asia, as a typical example, the total costs of charging, discharging and transportation (freight) can be in the order of US$ 20 to 30 per tonne of scrap. In other words, when the average scrap price say, the composite price is around US$ 125 to 135 per tonne of scrap in the United States, the delivered price in southern or eastern Asia could be between US$ 150 and 170 per tonne.

5.5 The different scrap categories

This is a complicated subject in that the "quality" of the various scrap categories is linked in two ways, upstream and downstream:

(a) Upstream, by the way the scrap is collected and prepared. Most classifications developed in different countries and detailed in the IISI study already mentioned are based on such a principle;

(b) Downstream, by the utilization intended for different scrap categories. As an example, to produce various grades of wire rod, the specifications and chemical composition required will be different.

It is here that the growing impact of "scrap substitutes" can be observed, particularly in the United States where Wulff[2] mentions the interaction of the price of domestic low residual scrap and imported pig iron prices (see figure 5.11). It is highly probable that there will be more and more such interactions in the future.

---

2 Gorham/Intech Conference, Charlotte, North Carolina, the United States, 3-5 March 1997.

Figure 5.11 Low residual scrap versus imported pig iron prices
(US$/gross tonne)

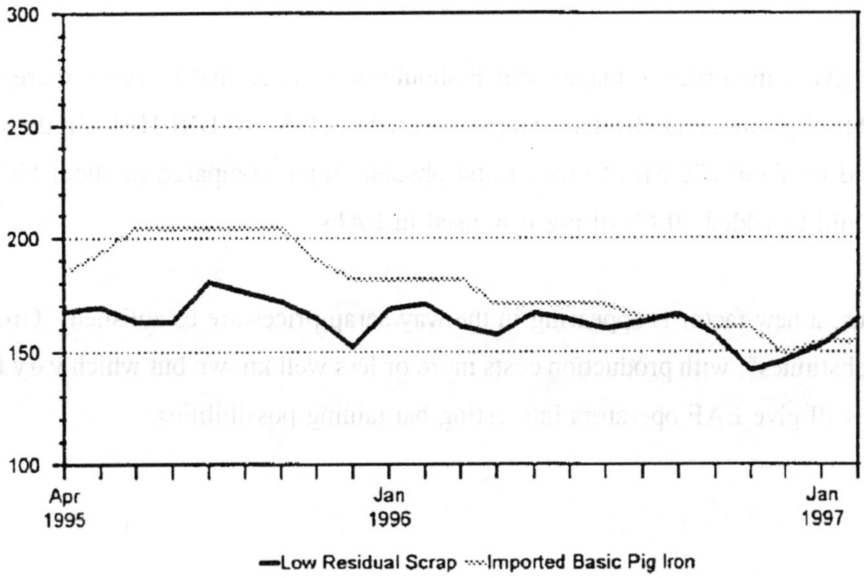

Source: David J. Joseph Company.

5.6 Conclusions

The first factor is that the scrap market and the way prices are established is mainly fixed by the growing demand of EAFs and mini mills, and the growing supply of obsolete scrap (the prompt scrap supply being more constant and often directly "allocated" outside the main market), which, itself is complicated from the primary scrap collectors to the scrap dealers and preparers with the stockyards and shredders.

The second factor is that in the past the market for scrap, mainly EAFs in mini mills was different from the market for iron ore or the primary metals produced from iron ore, practically only pig iron. Iron ore and pig iron were used in integrated iron and steel plants; mini mills were accounting for a growing proportion of long products while the integrated plants were concentrating on flat products. In this way, the markets and prices of scrap, on the one hand, and iron ore, on the other, had independent evolutions. The evolution of scrap prices was accordingly influenced by steel demand (see figures 5.7 and 5.8) but more particularly by demand for light long products as was seen in Europe a few years ago at a time when there was a huge demand for the export of such products to China.

Now, the mini mills are "invading" all types of product sectors especially in the United States and the competition is growing between collected scrap and the primary metals produced from iron ores, i.e. pig iron and DRI/HBI including iron carbide.

To give some order of magnitude, it should be recalled that in 1995, there was a supply of about 240 Mt of prompt and obsolete scrap compared to 30 Mt of DRI/HBI. In the year 2000, these figures could be about 270 Mt of prompt and obsolete scrap compared to about 50 Mt of DRI/HBI to which could be added 20 Mt of pig iron used in EAFs.

Thus, a new factor is appearing in the way scrap prices are established. Growing quantities of "scrap substitutes", with production costs more or less well known but which vary from one region to another, will give EAF operators interesting bargaining possibilities.

Table 5.1 United States monthly crude steel production and No. 1 HMS composite price

	Crude steel production (Mt)	Scrap prices in US$ per gross tonnes
1995		
January	8 092	141.58
February	7 449	138.67
March	8 284	133.58
April	7 755	134.67
May	7 903	137.67
June	7 517	134.65
July	7 537	133.17
August	7 670	143.12
September	7 864	140.92
October	7 879	138.58
November	7 779	133.33
December	7 873	136.00
1996		
January	8 147	143.23
February	7 655	142.75
March	8 288	135.64
April	7 791	132.76
May	7 982	138.76
June	7 857	136.40
July	7 788	132.33
August	7 826	134.18
September	7 627	136.23
October	7 895	127.43
November	7 508	115.14
December	7 883	121.98
1997		
January	7 925	127.44
February	7 499	134.04
March	8 550	128.88

Source: *American Metal Market.*

Table 5.2 Evolution of scrap import prices by countries

(US$/tonne)

Region		1990	1991	1992	1993	1994	1995	1996
ECE	Belgium	197.99	150.34	163.91	132.56	104.76	137.13	144.32
Region	Denmark	121.40	88.01	67.87	89.71	104.42	317.25	128.05
	France	152.22	110.11	93.76	104.59	148.63	172.00	-
	Germany *	199.36	208.17	198.00	170.40	105.85	133.12	-
	Ireland	132.33	112.27	98.85	130.15	170.93	222.65	-
	Italy	146.39	116.77	107.68	128.35	125.93	164.56	-
	Netherlands	164.45	126.49	108.84	123.92	130.21	132.66	154.35
	Norway	115.38	106.05	92.34	115.14	159.73	189.40	126.63
	Portugal	121.21	97.91	90.03	153.83	345.28	367.51	171.07
	Spain	152.60	129.31	128.30	133.31	105.45	148.54	179.78
	Switzerland	143.35	115.61	94.38	100.66	142.24	161.30	137.48
	Turkey	142.67	129.10	112.46	128.62	118.46	144.83	-
Non-	Argentina	-	-	108.87	114.45	86.97	148.66	-
ECE	Brazil	-	-	112.43	145.76	19.27	43.00	-
region	China	-	-	147.42	153.73	144.04	141.52	-
	Colombia	-	114.30	103.07	104.10	104.64	79.32	-
	Hong Kong	-	-	157.28	175.78	137.10	100.82	225.97
	Indonesia	154.10	141.42	125.53	154.84	234.66	1316.24	-
	Jordan	142.72	120.61	141.09	116.63	133.35	171.46	-
	Korea, R.O.	165.73	189.47	145.96	157.81	161.39	179.31	-
	Malaysia	160.27	151.93	129.17	154.72	217.38	231.62	-
	Pakistan	150.37	137.20	136.24	144.45	165.48	188.38	-
	Paraguay	74.40	-	-	204.26	-	105.60	-
	Philippines	-	160.42	140.81	167.13	202.40	299.28	-
	Singapore	141.35	126.99	114.86	141.08	101.83	94.49	126.16
	Venezuela	137.44	-	82.30	135.28	140.93	146.39	-

Source: United Nations Statistical Office, COMTRADE database.

* Data for 1990 are for the former West Germany.

CHAPTER 6

SCRAP SUBSTITUTES

"Scrap substitutes" or more precisely, the primary metals used, especially in EAFs, as a complement or supplement to scrap have already been mentioned several times in this study. Chapter 6 provides a clear definition of the different scrap substitutes, reviews the growing use of pig iron/hot metal in EAFs and the corresponding trade worldwide, examines, in a similar way, the growing use of DRI/HBI and iron carbide in EAFs, the local and national trade, and the growing international trade in these products, and will finally draws conclusions from this survey.

6.1 What are "scrap substitutes" ?

Scrap substitutes can be divided into two categories, based on their properties and on the way they are produced. The first category concerns the pig iron produced in blast furnaces or in smelting reduction processes like COREX by the reduction of iron ore and the melting of the metal. This can be used as liquid "hot metal" or in the solid form at room temperature as "pigs", "plates" or granulated iron. The second category concerns the products of direct reduction (DR), including iron carbide, which are produced in increasing amounts by the reduction of iron ore in the solid state. Usually, they are used cold like pig iron but new technology is introducing the use of hot DRI, i.e. not cooled after reduction and transported directly to the EAF.

It should be noted that most of the pig iron produced worldwide is used in integrated iron and steel plants in BOFs, and only a small but growing part is charged as a scrap substitute in EAFs. On the other hand, most DRI or HBI and iron carbide is used in EAFs.

Table 6.1 summarizes the properties of these various "scrap substitutes" and the way they are used. It can be clearly seen that all of them have similar Fe content, low metalloid and non-ferrous metal contents (the "residual" elements such as Cu, Co, Ni, Cr, etc.), but differ in the content of C, O and gangue.

Table 6.1 Chemical composition of scrap substitutes

Physical state	Hot liquid	Hot	COLD		
	Hot metal	DRI	DRI/HBI	Pig iron	IC
Temperature (°C)	1300	600/700	0/30	0/30	0/30
Chemical composition					
Fe%	93/96	92/94	92/94	93/96	89/91
C%	4	0/2	0/2	4	6/7
Ga...%	0	2/4	2/4	0	2/4
0%	0	1/2	1/2	0	0/1
S%	always low or very low				
P%	always low or very low				
Non-ferrous metals%	negligible				

Source: UN/ECE Working Party on Steel.

6.2 The importance of pig iron/hot metal as EAF charge

The evolution of world pig iron production was described in chapter 1, section 1.5. It should be recalled that most of it is used directly in the BOFs of conventional iron and steel plants. However, some is used in foundries and some in EAFs as a "scrap substitute". The latter use can be divided into two categories:

(a) Hot metal as produced by a nearby blast furnace. It is difficult to assess the exact quantities as this is a fairly recent practice and depends on the specific location. It can, however, be estimated from the operation of plants such as Usinor Sacilor at Gandrange, France; Cockerill Sambre at Charleroi, Belgium; Ferriere Nord at Osoppo, Italy, and a number of plants in Brazil, China, Japan, India, etc. The total is estimated at 3 million tonnes of hot metal per year. However, in view of future uncertainties, a more realistic estimate would be between a minimum of 2 million tonnes and possibly up to 5 million tonnes per year, if not more. Furthermore, the total is certainly increasing due to the need for scrap substitutes, especially in developing areas where new processes or combinations of processes, such as Corex-Midrex in Hanbo, in the Republic of Korea or Saldanha Bay in South Africa have led to the use of hot metal and DRI in EAFs. ISPAT at Raigad is associating Midrex and a blast furnace.

(b) Pig iron as used in EAFs, in solid state, from the same country (i.e. an EAF and BF in the same country) or imported from another country. In the latter case, statistical data exist which gives some idea of the volume utilized (see tables 6.2 and 6.3), but it should be remembered that not all these amounts of pig iron are used in EAFs but may also be shipped to foundries or BOFs.

The replies to the UN/ECE questionnaire provide more precise data about the use of pig iron in EAFs but they do not distinguish between use of hot metal and pig iron in EAFs. The replies are summarized in table 6.4.

Table 6.2 Trends in pig iron exports
(Thousands of tonnes)

Country or region	1980	1985	1990	1994
CIS	4500	4465	6276	4396
(Russian Federation)	-	-	-	(2699)
(Ukraine)	-	-	-	(1657)
Brazil	841	2476	3489	2631
China	-	4	549	1557
Japan	15	1085	25	1222
Western Europe	1868	1238	800	613
Eastern Europe	63	86	804	469
All others	1364	1341	1619	1499
Total	8651	10695	13562	12387

Source: IISI, Statistical Yearbooks.

Table 6.3 Trends in pig iron imports
(Thousands of tonnes)

Country or region	1980	1985	1990	1994
Western Europe	2407	1841	3563	3645
Eastern Europe	4238	4106	3870	191
United States	423	307	347	2441
Asia	1248	4153	6832	6310
(China)	(353)	(2911)	(1215)	(401)
(Japan)	(181)	(748)	(3263)	(1811)
All others	155	409	219	312
Total	8471	10806	14841	12899

Source: IISI, Statistical Yearbooks.

Table 6.4 Consumption of pig iron in EAFs
(Thousands of tonnes)

Country or region	1995	1996	1997 (Predictions)
United States	1200	1900	-
EU (Incomplete)	1072	-	-
Russian Federation	368	370	400
Japan	3060	2912	-
Korea, Rep. of	1460	2050	2460
Total	**7160**	-	-

Source: Governments replies to the UN/ECE questionnaire.

From this information it is clear that the world trade in pig iron has increased from 8.5 Mt to a level of about 13 Mt in 15 years as shown in figure 6.1. The main importer, Asia, is followed by western Europe and the United States while the main exporters are the CIS (Russian Federation and Ukraine) and Brazil although China and Japan have started becoming exporters. The total world use of pig iron as a "scrap substitute" for EAFs, is according to the data shown in table 6.4, at least 7 Mt in 1995 and probably around 10Mt in 1997.

Figure 6.1 Schematic map of international maritime trade of pig iron in 1996

Source: UN/ECE Working Party on Steel.

6.3 The development of the production and trade of DRI/HBI as scrap substitutes

The development of the production of DRI/HBI was discussed in chapter 1. Unlike pig iron, practically all DR production is used in EAFs. Only a small amount, probably less than 1 Mt per year out of a total of 33 Mt (in 1996) is used in blast furnaces (in the United States) or in foundries.

This growing use of DRI/HBI in EAFs is mainly in plants where EAFs and DR facilities are integrated. Most of them are in developing areas and correspond exactly to the idea of "scrap substitutes".

This maritime trade in DRI/HBI has developed as indicated in table 6.5. It is supplied by:

(a) The three merchant DR plants:

Fior (1976) (Venezuela)	0.4 Mt/y
Amsteel Labuan (1984) (Malaysia)	0.65 Mt/y
Opco (1990) (Venezuela)	1.00 Mt/y
Grasim (1993) (India)	0.75 Mt/y

This makes a total of about 2.8 Mt/y to which should be added iron carbide (0.15 Mt in 1996).

(b) DR plants feeding partly local steel plants such as:

Caribbean Ispat (1980-1982) up to 0.80 Mt/y

Venprecar (1990) (Venezuela)	up to 0.66 Mt/y
Essar (1990-1992) (India)	up to 1.32 Mt/y
Ispat Raigad (1994) (India)	up to 1.00 Mt/y
OEMK (1983-1988) (Russian Federation)	up to 1.67 Mt/y

This amounts to a total of between 0.6 to 2.00 Mt/y (excess to be exported).

(c) New merchant plants being built such as:

Finmet BHP, Australia (1998)	2.5 Mt/y
Circored Cliffs, Trinidad (1998)	0.5 Mt/y
Midrex Comsigua, Venezuela (1996)	1.0 Mt/y
Hyl III Posven, Venezuela (1999)	1.5 Mt/y
Midrex Ispat, Trinidad (1999)	1.3 Mt/y

Other future plants could include:

Tondu Corp. (mainly from Canada)	0.9 Mt/y
Finmet OIC Venezuela	2.5 Mt/y

Table 6.5 Estimate of the maritime DRI/HBI/IC trade
(Millions of tonnes)

Year	Trade
1990	1.79
1991	2.25
1992	2.24
1993	2.90
1994	3.46
1995	3.76
1996	3.40
1997	4
1998	4
1999	7
2000	10
2001	15

Source: UN/ECE Working Party on Steel, Midrex Direct Reduction Corporation.

From these data and estimates, the total maritime trade in DRI, HBI and iron carbide is summarized in table 6.6 and figure 6.2.

Table 6.6 Schematic structure of DRI/HBI/IC maritime trade
(Thousands of tonnes)

Country or region	1993	1994	1995	**1996**
Main exporters				
Venezuela	1000	1000	1500	1500
Malaysia	723	829	900	1000
Russian Federation	332	405	500	400
Trinidad & Tobago	130	250	200	300
India	100	450	350	200
All others	300	366	300	200
Total	2585	3300	3750	3600
Main importers				
United States	1060	1170	1190	1181
Asia	1240	2180	2370	1919
Western Europe (Turkey included)	100	150	200	300
Total	2400	3500	3760	3400

Source: UN/ECE Working Party on Steel, United States Geological Survey.

Figure 6.2 Schematic map of international maritime trade in DRI, HBI and IC in 1996
(Millions of tonnes)

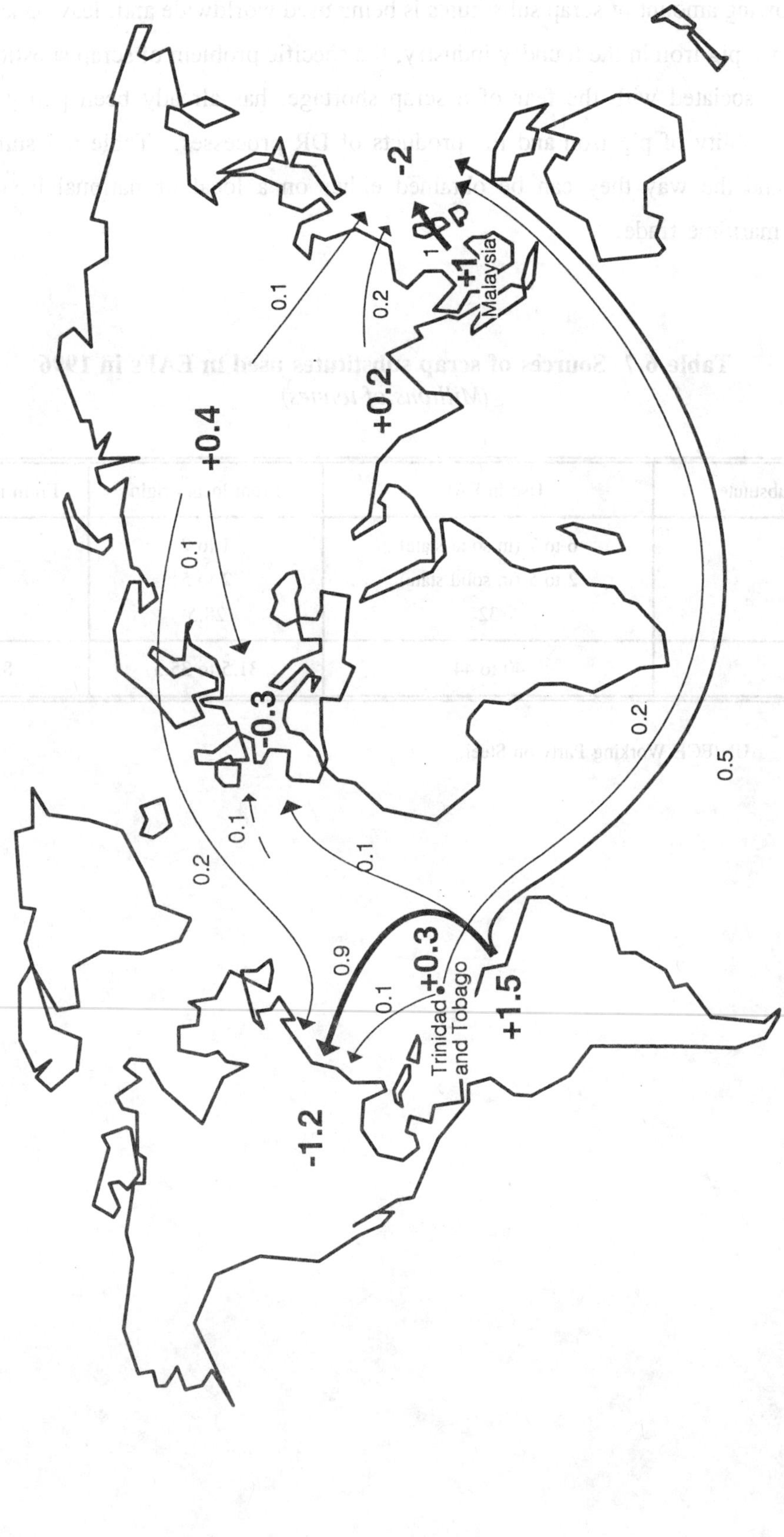

Source: UN/ECE Working Party on Steel.

6.4 Conclusions about the growing importance of scrap substitutes

A growing amount of scrap substitutes is being used worldwide and, leaving aside the special case of scrap vs. pig iron in the foundry industry, the specific problem of scrap substitutes in electric steelmaking, associated with the fear of a scrap shortage, has already been partly solved by the increasing availablity of pig iron and the products of DR processes. Table 6.7 summarizes these possibilities and the way they can be obtained either on a local or national basis or from the international maritime trade.

Table 6.7 Sources of scrap substitutes used in EAFs in 1996
(Millions of tonnes)

Scrap substitute	Use in EAFs	From local origin	From maritime trade
Pig iron	6 to 7 (in solid state)	1 to 2	5 to 6
Hot metal	2 to 5 (in solid state)	2 to 5	0
DRI/HBI/IC	32	28.5	3.5
Total	40 to 44	31.5 to 35.5	8.5 to 9.5

Source: UN/ECE Working Party on Steel.

CHAPTER 7

COUNTRY COMMENTARIES

The world can be divided into three categories of countries —"industrialized", "transition" and "developing"— on the basis of the data collected, using the criteria of scrap resources, scrap utilization and industrial structure.

7.1 Industrialized countries

The industrialized countries comprise four main regions of the world, with different scrap patterns:

(a) North America, which is still the largest scrap-exporting region and at the same time is becoming a very important importer of "scrap substitute". Furthermore, it is also becoming an important producer of DRI/HBI;

(b) Western Europe (i.e. the European Union, Norway, Switzerland and Turkey), whose scrap imports and exports are more or less balanced. The countries of the north are important exporters while those of the south are the main importers. DR does not yet play an important role in the region, but a certain amount of pig iron is used as a scrap substitute in EAFs, mainly for quality control;

(c) Japan, which is in a similar situation to western Europe regarding scrap and scrap substitutes;

(d) Oceania, i.e. mainly Australia which is exporting less and less scrap but will probably become am important DRI/HBI exporter.

The following general comments should be made regarding these four regions which have many similarities in that they are more or less self sufficient regarding their scrap supply or, possibly small net exporters or importers. In the future, it seems that all of them will be net exporters.

North America, particularly the United States, saw some impressive developments. For example, the widescale development of electric steelmaking will represent 50% of steel production in the near future. On the one hand, considerable amounts of scrap will be exported taking into account the fact that North America has been the largest long-time scrap exporter. On the other hand, import demand for scrap including a considerable amount of scrap substitutes will continue to increase, which will start off the production of DRI and IC.

Western Europe, with the exception of Turkey, is more or less self sufficient in terms of scrap supply and will quite probably become an exporter with limited tonnages (see table 7.1). Table 7.1 again indicates the difficulty in obtaining precise scrap statistical data. However, the order of magnitude and the trends are clear.

Japan is now self-sufficient in scrap supply (see chapter 4, table 4.5) and could in the future become a scrap exporter.

Australia has traditionally been a small scrap exporter, practically only to Asia, but with the development of mini mills, the volume of scrap exported has shown a decreasing trend. Although, as indicated in table 7.2, there are discrepancies between the various sources, and fluctuations from year to year, there is a general decreasing trend. In spite of the discrepancies, it can be seen that Australia is and will probably remain a small scrap exporter. It should also be noted that Australia will become an important HBI exporter with the commissioning next year of BHP's Finmet plant at Port Hedland with a capacity of 2.2 Mt/y to which should be added a number of similar projects in western and southern Australia.

Table 7.1 Net scrap trade of EU

(Thousands of tonnes)

Year	Eurostat	IISI	UNCTAD
1990	-395	-327	-1049
1991	-	+1485	+1720
1992	-	+2373	+2350
1993	-	+4408	+4903
1994	-	+377	+1380
1995	+1882		-878
1996	+1644	-	-

Source: Eurostat Statistics, IISI Statistical Yearbook, UNCTAD, Handbook of World Material Trade Statistics, 1990-1995.

Table 7.2 Scrap exports from Australia
(Thousands of tonnes)

Year	UN/ECE study	IISI	UNCTAD
1975	32.9	734	-
1980	94.2	708	-
1981	604	498	-
1982	492	554	-
1983	424	588	-
1984	500	445	-
1985	379	379	-
1986	536	601	-
1987	922	922	-
1988	1154	1139	-
1989	1154	1496	-
1990	1301	1922	220
1991	1353	1185	278
1992	955	1035	388
1993	759	758	766
1994	398	399	403
1995	309	-	309
1996	680	-	-

Source: UN/ECE Working Party on Steel, IISI Statistical Yearbook, UNCTAD, Handbook of World Material Trade Statistics, 1990-1995.

7.2 Transition countries

The transition countries can to be divided into three groups: eastern and central Europe, the Baltic States and the Commonwealth of Independent States.

(a) The eastern and central European countries have completely changed their position in that they are now scrap exporters and no longer pig iron importers. This situation could change in the future as, after the switch from open hearth steelmaking and ingot casting to oxygen steelmaking and continuous casting, they will probably develop electric steelmaking and, as a consequence, export less scrap, and begin to import or produce some scrap substitutes such as pig iron or HBI.

(b) The three Baltic States have practically no ironmaking facilities but open hearth steelmaking based on scrap of Latvia of about 280,000 t/y, small scale electric steelmaking also based on scrap in Estonia (3,000 t/y) and small scale induction furnaces also based on scrap in Lithuania. In any case, these three countries are scrap exporters and even if some discrepancies exist in the data, they probably export together between 700,000 and 800,000 tonnes of scrap per year.

(c) In the CIS similar technological trends will probably appear but at a slower pace than in the eastern European countries. At present, the various countries of the CIS and, especially the Russian Federation and Ukraine are important scrap exporters; very important pig iron exporters; and a DRI exporters from the Koursk (OEMK) DR plant in the Russian Federation to which will be added a new DR plant at the Lebedinsky mine.

7.3 Developing countries

These countries must be split into the following categories: Latin America, Africa, the Middle East and eastern and southern Asia.

Latin America is developing and, on the whole, is a small scrap importer but the scrap substitutes have already begun to play a role as shown in chapters 1 and 6. These substitutes, mainly pig iron in Brazil where it is used both as hot metal or pig irons in Brazilian EAFs to substitute scrap or exported, as seen in chapter 6 (Brazil is, after the CIS, a very important pig iron exporter), and DRI/HBI which is mainly produced. In Mexico and Argentina the scrap substitute is used to feed EAFs. In Venezuela and Trinidad and Tobago, part of their scrap substitute production is used in the local steel plants (all of them based on EAFs) of Sidor, Sivensa and Caribbean ISPAT but a growing part is exported Venezuela is and will probably remain the world's top DRI/HBI exporter.

Africa is, as is well known, not developing very fast except in Northern and Southern Africa: the steel industry in the Republic of South Africa is developing mainly from its important coal resources. In Saldanha Bay, a combination of Corex (the second after the Corex unit of ISCOR's Pretoria works, which was the first worldwide) and Midrex is being constructed. The EAF steelmaking plants in South Africa use local scrap and DRI is produced in coal-based, rotary-kiln DR plants. There are a number of gas-based DR projects supplied mainly to export HBI. In North Africa, there are several projects for new gas-based DR plants associated with EAF, especially in Egypt, but as in other African countries, the developments are slow, at least for the time being. In Africa, generally speaking, with the exception of South Africa, countries are not participating much in scrap or scrap substitutes trade. In any case, the whole continent is a small scrap exporter.

The Middle East with the exception of the Islamic Republic of Iran which operates, among other DR based plants, a conventional iron and steel plant, over all its steel developments to a combination of gas-based DR and EAFs. As a consequence, the region is a small scrap exporter and could be, in the future, an exporter of DRI or HBI or iron carbide. However, for the time being, there is no export-oriented DR production.

Eastern and southern Asia is the region which has seen by far the largest expansion of steel consumption and production. This has led to a special situation where there is not enough scrap and the region (see chapter 4, especially figure 4.3) is the main scrap importing area in the world. However, as indicated in chapter 2, especially in table 2.11, the scrap collection is now growing with the economic development of these countries.

Furthermore, with difficult scrap supply and high scrap prices, the production or import of substitutes has been developing, i.e. both DRI/HBI (see chapter 1, section 5) and pig iron. There are all types of development in this field, such as conventional DR processes based, depending on local resources, on coal (China, India, Thailand) or gas (India, Malaysia, Indonesia) and pig iron production with the conventional blast furnace process (China, Taiwan Province of China, Republic of Korea and now Thailand) and, also mini blast furnaces (especially in China but also in India) as well as through new processes like Corex in the Republic of Korea and India.

CHAPTER 8

CONCLUSIONS

Although most of the figures provided by countries in response to the questionnaire are precise, many do not always correspond to the same definitions. Some data may also be missing, which explains, for example, the systematic discrepancy each year between total world imports and exports (see as an example, chapter 4, tables 4.14 and 4.15). However, the orders of magnitude and trends are accurate and very clear. Concerning the precision of world figures: the margin of error is most probably an underestimation rather than an overestimation. For example, world demand (table 3.16) and supply (table 2.13) for 1966, without scrap substitutes, is probably closer to 400 Mt than the 381.5 or 395 Mt indicated and international scrap trade (tables 4.14 and 4.15 for 1995 and 1994) is probably closer to the highest imports figures, i.e. 46 Mt in 1994 and 51 Mt in 1995.

The first conclusion (see chapter 1) is that scrap demand is increasing mainly due to (see chapter 3) the development of electric steelmaking. Table 3.16 clearly indicates the volume of scrap consumption (scrap and scrap substitutes) in electric steelmaking which can be estimated in 1996 at 266 Mt out of a total of 421 Mt or 63%. All other demand is either more or less stable or declining especially for open hearth furnaces. The demand from EAFs will certainly increase and could move in the years 2000 and 2005 to the levels indicated in table 3.1.

Table 8.1 Estimated evolution of total scrap demand worldwide
(Millions of tonnes)

	1996	2000	2005
Foundry	47	47	47
Blast furnace	3	3	3
Open hearth steelmaking	25	20	10
Oxygen steelmaking	80	77	75
Electric steelmaking	265	283	305
Total	420	430	440

Source: Data from chapter 3 and for estimates UN/ECE Working Party on Steel.

The second conclusion is that regarding scrap **supply** (see chapter 2), the following two categories have to be separated. First, circulating scrap which is usually directly recycled where it is produced and only traded in small amounts (the amounts of circulating scrap are continuously decreasing). Secondly, the prompt and particularly the obsolete scrap which is collected and purchased.

In this way, the picture given in table 3.16 can be completed to give the complete worldwide scrap balance for 1996 (table 8.2).

Table 8.2 Total estimated worldwide scrap balance for 1996

(Millions of tonnes)

Demand		Supply	
Foundry	47	Circulating scrap	
Blast furnaces	3	Iron and steel	100
Steelmaking		Foundry	20
Open hearth	25	Plants capital scrap	5
Oxygen	80	Prompt scrap	100-120
Electric	265	Obsolete scrap	135-170
Sub-total	420	Sub-total	380-395
Total	420	**Total**	420

The third conclusion is that the importance of what are called **scrap substitutes** (see chapter 6) is growing. The total amount of scrap substitutes was estimated at 35 to 40 Mt in 1996 but as seen in chapter 6 is probably closer to 40 to 44 Mt by the year 1996, and may reach 60 or even 75 Mt by the year 2000. Such estimates should dissipate any fear of a scrap shortage and, on the contrary, point to a potential oversupply of "metallics" or "primary metals", at the beginning of the next century.

The fourth conclusion relates to **prices** (see chapter 5). The study has shown that there is a world scrap market and that, expressed in the same currency, say in US$, the level of prices is more or less the same at any given time in all the industrialized countries. However, in developing countries where scrap has to be imported over long distances, the level of prices is higher due to transportation costs. In such a situation there is a strong incentive to produce scrap substitutes or even to return to the conventional iron and steel plant concept, as is happening in a number of Asian countries.

In addition, there has been an unexpected stagnation in the maritime scrap trade. However, the trade in DRI/HBI and even in pig iron seems more active.

Finally, the influence of quality is important, both of steel products and of scrap and scrap substitutes (see below, about the United States). This will probably lead to more acute competition between the various categories of scrap (especially higher grade prompt scrap) and the "substitutes", DRI/HBI, iron carbide and pig iron.

The fifth conclusion concerns the **trade** within a country or a region compared with international maritime trade (see chapter 4). It is clear enough that the industrialized countries have now become self-sufficient or exporters This is the case in western Europe and now eastern Europe, Japan and Australia.

The United States in particular is very interesting in that it remains the largest scrap exporter while also being the main DRI/HBI importer, and is beginning to produce DRI and iron carbide. Here, the important aspect of the quality of scrap and scrap substitutes becomes clear, as new EAF-based steel plants move "upmarket" to produce higher steel grades of long and now flat products.

Furthermore, figure 4.3 indicates the main flow of international maritime scrap trade. This trade of about 20 Mt per year is relatively small compared with the total scrap trade of about 250 to 300 Mt per year, or the total international scrap trade of about 45 to 50 Mt per year which is, thus, mainly regional.

The maps in figures 6.1 and 6.2 and tables 8.1 and 8.2 give the corresponding information for pig iron and DRI/HBI/IC trade.

ANNEX 1

TOTAL SCRAP BALANCE BY COUNTRY

Other countries*

* *Countries for which insufficient data were available to calculate the scrap balance.*

AUSTRIA

(Thousands of tonnes)

	1980	1985	1990	1993	1994	1995	1996
Basic Iron and Steel Data							
1. Pig iron production	3 485	3 734	3 452	3 070	3 320	3 878	3 416
2. DRI production	...	...	...	...	...	...	...
3. Crude steel production	4 624	4 661	4 291	4 149	4 399	4 990	4 442
4. CC ratio (%)	51	93	96	96	96	96	97
Iron and Steel Scrap Data							
I. CONSUMPTION (total)	1 694	1 525	1 236	1 436	1 442	1 659	1 511
1. Iron and steel works	1 694	1 525	1 236	1 436	1 442	1 659	1 511
2. Steel foundries	...	...	...	...	...	...	...
3. Iron foundries	...	...	...	...	...	...	...
4. Re-rolling	...	...	...	...	...	...	...
5. Independent electric ferro-alloy	...	...	...	...	...	...	...
II. OWN RESOURCES (total)	833	581	628	548	566	524	482
1. Circulating scrap	833	581	628	548	566	524	482
(a) Iron and steel works	833	581	628	548	566	524	482
(b) Steel foundries (independent)	...	...	...	...	...	...	...
(c) Iron foundries	...	...	...	...	...	...	...
2. Capital scrap (demolition, etc.)	-	-	...	...	...	...	...
III. EXTERNAL RECEIPTS (total)	861	865	514	903	915	1 151	966
1. Iron and steel works	861	865	...	...	...	1 151	966
(a) Trade	861	865	...	...	...	...	...
(b) Direct purchase	...	...	...	...	...	...	...
(c) Imports (not included in (a) or (b))	...	...	...	...	...	...	...
2. Steel foundries (independent)	...	...	...	...	...	...	...
3. Iron foundries	...	...	...	...	...	...	...
4. Independent electric ferro-alloy	...	...	...	...	...	...	...
IV. OUTSIDE SALES (-) (total)	...	...	...	...	...	4	5
1. Iron and steel works	...	...	...	...	...	4	5
2. Steel foundries (independent)	...	...	...	...	...	...	...
3. Iron foundries	...	...	...	...	...	...	...
V. STOCK VARIATION (total)	-	-79	...	...	...	12	-69
1. Iron and steel works	-	-79	...	...	...	...	...
2. Steel foundries (independent)	...	...	...	...	...	...	...
VI. FOREIGN TRADE (balance)	-130	-207	288	43	59	-40	...
1. Imports (-)	-143	-239	-55	-418	-408	-529	...
2. Exports (+)	13	32	343	460	467	489	...
A. TOTAL NATIONAL RESOURCES (utilized)(II + III + VI)	1 564	1 239	1 430	1 494	1 540	1 635	...
B. DOMESTIC SCRAP (total) (A - II.1)(collection utilized)	731	658	802	946	974	1 111	...
C. DOMESTIC SCRAP MARKETED (B - II.2) or (A - II)	731	658	802	946	974	1 111	...
D. EXTERNAL PURCHASE REQUIREMENTS (GROSS)(I - II) (all consumers)	861	944	608	888	876	1135	1 029
E. EXTERNAL PURCHASE REQUIREMENTS (NET) (I - II + IV) (all consumers)	861	944	...	...	...	1139	1 034
F. SURPLUS (+) or DEFICIT (-) (balance) (C - E) or (VI + V)	-130	-286	...	...	...	-27	...

Source: Government replies to UN/ECE questionnaires.

Note: Stock variation calculated as follows: V = II + III - I - IV.

AZERBAIJAN

(Thousands of tonnes)

	1980	1985	1990	1993	1994	1995	1996	1997 Forecast
Basic Iron and Steel Data								
1. Pig iron production	...	...	...	...	...	...		
2. DRI production	...	...	...	...	...	...		...
3. Crude steel production	...	...	677	236	40	21	1	153
4. CC ratio (%)	...	...	...	...	8	5	...	
Iron and Steel Scrap Data								
I. CONSUMPTION (total)	...	...	473	...	...	25	1	127
1. Iron and steel works	...	...	473	...	...	25	1	127
2. Steel foundries	...	...	...	...	...	...		...
3. Iron foundries	...	...	...	...	...	...		...
4. Re-rolling	...	...	...	...	...	...	...	
5. Independent electric ferro-alloy	...	...	...	...	...	...	...	
II. OWN RESOURCES (total)	...	...	14	...	...	2	1	2
1. Circulating scrap	...	...	14	...	...	2	1	2
(a) Iron and steel works	...	...	14	...	...	2	1	2
(b) Steel foundries (independent)	...	...	...	...	...	...		...
(c) Iron foundries	...	...	...	...	...	...		...
2. Capital scrap (demolition, etc.)	...	...	0	...	...	-	-	-
III. EXTERNAL RECEIPTS (total)	...	...	2	...	...	-	-	-
1. Iron and steel works	...	...	2	...	...	-	-	-
(a) Trade	...	...	...	...	...	-	-	-
(b) Direct purchase	...	...	...	...	...	-	-	-
(c) Imports (not included in (a) or (b))	...	...	...	...	...	-	-	-
2. Steel foundries (independent)	...	...	...	...	...			
3. Iron foundries	...	...	...	...	...	...	...	
4. Independent electric ferro-alloy	...	...	...	...	...		...	
IV. OUTSIDE SALES (-) (total)	...	...	...	...	...		...	
1. Iron and steel works	...	...	...	...	...	...	...	
2. Steel foundries (independent)	...	...	...	...	...	...	...	
3. Iron foundries	...	...	...	...	...	...	...	
V. STOCK VARIATION (total)	...	...	...	...	...		...	
1. Iron and steel works	...	...	...	...	...	...	...	
2. Steel foundries (independent)	...	...	...	...	...		...	
VI. FOREIGN TRADE (balance)	...	...	...	5	11	12	12	...
1. Imports (-)	...	...	...	-	-	...	...	
2. Exports (+)	...	...	...	5	11	12	12	...
A. TOTAL NATIONAL RESOURCES (utilized)(II + III + VI)	...	...	...	...	...	14	13	...
B. DOMESTIC SCRAP (total) (A - II.1)(collection utilized)	...	...	...	...	...	12	12	...
C. DOMESTIC SCRAP MARKETED (B - II.2) or (A - II)	...	...	...	...	...	12	12	...
D. EXTERNAL PURCHASE REQUIREMENTS (GROSS)(I - II) (all consumers)	...	...	459	...	...	23	0	125
E. EXTERNAL PURCHASE REQUIREMENTS (NET) (I - II + IV) (all consumers)	...	...	...	...	...	...	...	...
F. SURPLUS (+) or DEFICIT (-) (balance) (C - E) or (VI + V)	...	...	...	...	...	...	...	...

Source: Government replies to UN/ECE questionnaires.

BELARUS

(Thousands of tonnes)

	1980	1985	1990	1993	1994	1995	1996
Basic Iron and Steel Data							
1. Pig iron production	...	...	...	...	...	...	...
2. DRI production	...	...	...	...	...	...	...
3. Crude steel production	...	...	1 112	946	880	744	886
4. CC ratio (%)	...	...	...	...	...	-	-
Iron and Steel Scrap Data							
I. CONSUMPTION (total)	...	...	715	1 007	911	636	859
1. Iron and steel works	...	...	715	1 007	911	636	859
2. Steel foundries	...	...	...	...	...	...	...
3. Iron foundries	...	...	...	...	...	...	...
4. Re-rolling	...	...	...	...	...	...	...
5. Independent electric ferro-alloy	...	...	...	...	...	...	...
II. OWN RESOURCES (total)	...	...	84	107	70	76	96
1. Circulating scrap	...	...	73	107	70	64	82
(a) Iron and steel works	...	...	73	107	70	64	82
(b) Steel foundries (independent)	...	...	...	...	...	...	...
(c) Iron foundries	...	...	...	...	...	...	...
2. Capital scrap (demolition, etc.)	...	...	11	...	...	12	14
III. EXTERNAL RECEIPTS (total)	...	...	612	771	735	545	760
1. Iron and steel works	...	...	612	771	735	545	760
(a) Trade	...	...	612	601	488	309	392
(b) Direct purchase	...	...	...	170	10	13	-
(c) Imports (not included in (a) or (b))	...	...	...	-	237	223	368
2. Steel foundries (independent)	...	...	...	...	...	...	...
3. Iron foundries	...	...	...	...	...	...	...
4. Independent electric ferro-alloy	...	...	...	...	...	...	...
IV. OUTSIDE SALES (-) (total)	...	...	...	...	...	...	...
1. Iron and steel works	...	...	...	...	...	...	...
2. Steel foundries (independent)	...	...	...	...	...	...	...
3. Iron foundries	...	...	...	...	...	...	...
V. STOCK VARIATION (total)	...	...	...	...	...	...	...
1. Iron and steel works	...	...	40	10	- 12	...	...
2. Steel foundries (independent)	...	...	...	...	...	...	...
VI. FOREIGN TRADE (balance)	...	...	30	256	179	20	-223
1. Imports (-)	...	...	-	- 32	- 11	-234	-401
2. Exports (+)	...	...	30	288	190	254	178
A. TOTAL NATIONAL RESOURCES (utilized)(II + III + VI)	...	...	726	1 134	984	640	634
B. DOMESTIC SCRAP (total) (A - II.1)(collection utilized)	...	...	653	1 027	914	576	551
C. DOMESTIC SCRAP MARKETED (B - II.2) or (A - II)	...	...	642	1 027	914	565	537
D. EXTERNAL PURCHASE REQUIREMENTS (GROSS)(I - II) (all consumers)	...	...	632	900	841	560	762
E. EXTERNAL PURCHASE REQUIREMENTS (NET) (I - II + IV) (all consumers)	...	...	...	...	...	...	...
F. SURPLUS (+) or DEFICIT (-) (balance) (C - E) or (VI + V)	...	...	...	...	...	...	...

Source: Government replies to UN/ECE questionnaires.

BELGIUM

(Thousands of tonnes)

	1980	1985	1990	1993	1994	1995	1996
Basic Iron and Steel Data							
1. Pig iron production	9 845	8 719	9 416	8 179	8 979	9 199	8 627
2. DRI production	...	...	...	...	...	...	...
3. Crude steel production	12 322	10 683	11 360	10 119	11 265	11 540	10 752
4. CC ratio (%)	26	60	92	95	95	98	98
Iron and Steel Scrap Data							
I. CONSUMPTION (total)	3 814	3 033	3 065	2 983	3 337	3 535	2 950
1. Iron and steel works	3 647	3 033	3 065	2 983	3 337	3 535	2 950
2. Steel foundries	63	...	...	...	...	...	...
3. Iron foundries	104	...	...	...	...	...	...
4. Re-rolling	...	...	...	...	...	...	...
5. Independent electric ferro-alloy	...	...	...	...	...	...	...
II. OWN RESOURCES (total)	2 588	1 872	1 371	1 265	1 357	1 311	1 209
1. Circulating scrap	2 167	1 596	1 173	1 181	1 196	1 156	1 100
(a) Iron and steel works	2 106	1 596	1 173	1 181	1 196	1 156	1 100
(b) Steel foundries (independent)	21	...	...	...	...	...	...
(c) Iron foundries	40	...	...	...	...	...	...
2. Capital scrap (demolition, etc.)	421	276	198	84	161	155	109
III. EXTERNAL RECEIPTS (tota	1 488	1 374	1 831	1 868	2 198	2 409	2 193
1. Iron and steel works	1 382	1 324	1 831	1 868	2 198	2 409	2 193
(a) Trade	1 108	1 074	1 276	1 306	1 507	1 616	1 513
(b) Direct purchase	253	10	10	21	18	55	20
(c) Imports (not included in (a) or (b	21	240	545	541	673	738	660
2. Steel foundries (independent)	42	...	...	...	...	...	...
3. Iron foundries	64	50	...	...	...	...	...
4. Independent electric ferro-alloy	...	...	...	...	...	...	...
IV. OUTSIDE SALES (-) (total)	262	142	60	33	35	33	29
1. Iron and steel works	262	142	60	33	35	33	29
2. Steel foundries (independent)	...	...	...	...	...	...	...
3. Iron foundries	...	...	...	...	...	...	...
V. STOCK VARIATION (total)	-	71	77	117	183	152	423
1. Iron and steel works	-	21	- 21	14	- 48	17	26
2. Steel foundries (independent)	-	...	...	...	...	...	...
VI. FOREIGN TRADE (balance)	91	- 754	-1 016	- 821	-1 404	-1 769	-2 070
1. Imports (-)	- 435	-1 490	-1 885	-2 430	-2 992	-3 745	-3 811
2. Exports (+)	526	736	869	1 609	1 588	1 976	1 741
A. TOTAL NATIONAL RESOURCES							
(utilized)(II + III + VI)	4 167	2 492	2 186	2 312	2 151	1 951	1 332
B. DOMESTIC SCRAP (total)							
(A - II.1)(collection utilized)	2 000	896	1 013	1 131	955	795	232
C. DOMESTIC SCRAP MARKETED							
(B - II.2) or (A - II)	1 579	620	815	1 047	794	640	123
D. EXTERNAL PURCHASE REQUIREMENTS							
(GROSS)(I - II) (all consumers)	1 226	1 161	1 694	1 718	1 980	2 224	1 741
E. EXTERNAL PURCHASE REQUIREMENTS							
(NET) (I - II + IV) (all consumers)	1 488	1 303	1 754	1 751	2 015	2 257	1 770
F. SURPLUS (+) or DEFICIT (-) (balance)							
(C - E) or (VI + V)	91	- 683	- 939	- 704	-1 221	-1 617	-1 647

Source: Government replies to UN/ECE questionnaires.

Notes: Stock variation calculated as follows: V = II + III - I - IV.

1980 and 1985 pig iron data also include blast furnace ferro-alloys.

BRAZIL

(Thousands of tonnes)

	1980	1985	1990	1993	1994	1995	1996
Basic Iron and Steel Data							
1. Pig iron production	12 685*	18 961*	21 008	23 982	25 177	25 021	23 978
2. DRI production	...	285*	167	187	133	57	...
3. Crude steel production	15 337*	20 456*	20 567	25 207	25 747	25 076	25 237
4. CC ratio (%)	33*	44*	58	57	59	63	72
Iron and Steel Scrap Data							
I. CONSUMPTION (total)	6 468	6 998	6 574	6 761	6 554	6 371	7 460
1. Iron and steel works	6 468	6 998	6 574	6 761	6 554	6 371	7 460
2. Steel foundries	...	...	...	...	...	...	...
3. Iron foundries	...	...	...	...	...	...	...
4. Re-rolling	...	...	...	...	...	...	...
5. Independent electric ferro-alloy	...	...	...	...	...	...	...
II. OWN RESOURCES (total)	3 221	3 629	3 248	3 660	3 479	3 398	4 061
1. Circulating scrap	3 221	3 629	3 248	3 660	3 479	3 398	4 061
(a) Iron and steel works	3 221	3 629	3 248	3 660	3 479	3 398	4 061
(b) Steel foundries (independent)	...	...	...	...	...	...	...
(c) Iron foundries	...	...	...	...	...	...	...
2. Capital scrap (demolition, etc.)	...	...	...	...	...	...	...
III. EXTERNAL RECEIPTS (total)	3 327	3 527	2 878	2 755	2 900	3 242	3 441
1. Iron and steel works	3 327	3 527	2 878	2 755	2 900	3 242	3 441
(a) Trade	3 305	3 423	...	...	...	...	...
(b) Direct purchase	...	...	...	...	...	...	...
(c) Imports (not included in (a) or (b))	22	104	113	114	149	8	8
2. Steel foundries (independent)	...	...	...	...	...	...	...
3. Iron foundries	...	...	...	...	...	...	...
4. Independent electric ferro-alloy	...	...	...	...	...	...	...
IV. OUTSIDE SALES (-) (total)	...	...	...	...	...	...	...
1. Iron and steel works	...	...	...	...	...	...	...
2. Steel foundries (independent)	...	...	...	...	...	...	...
3. Iron foundries	...	...	...	...	...	...	...
V. STOCK VARIATION (total)	...	...	...	...	...	...	...
1. Iron and steel works	80	158	...	...	...	...	...
2. Steel foundries (independent)	...	...	...	...	...	...	...
VI. FOREIGN TRADE (balance)	- 22	- 104	- 113	- 99	- 148	-5	4
1. Imports (-)	- 22	- 104	- 113	- 114	- 149	-8	-8
2. Exports (+)	-	-	-	15	1	3	12
A. TOTAL NATIONAL RESOURCES (utilized)(II + III + VI)	6 526	7 052	6 013	6 316	6 231	6 635	7 506
B. DOMESTIC SCRAP (total) (A - II.1)(collection utilized)	...	...	...	...	...	...	...
C. DOMESTIC SCRAP MARKETED (B - II.2) or (A - II)	3 305	3 423	2 765	2 656	2 752	3 237	3 445
D. EXTERNAL PURCHASE REQUIREMENTS (GROSS)(I - II) (all consumers)	3 247	3 369	3 326	3 101	3 075	2 973	3 399
E. EXTERNAL PURCHASE REQUIREMENTS (NET) (I - II + IV) (all consumers)	...	...	...	...	...	...	...
F. SURPLUS (+) or DEFICIT (-) (balance) (C - E) or (VI + V)	...	...	...	...	...	...	...

Sources: Government replies to UN/ECE questionnaires; * International Iron and Steel Institute.

Note: Stock variation calculated as follows: V = II + III - I - IV.

BULGARIA

(Thousands of tonnes)

	1980	1985	1990	1993	1994	1995	1996	1997 Forecast
Basic Iron and Steel Data								
1. Pig iron production	1 539*	1 712*	1 141	1 013	1 470	1 607	1 504	1 480
2. DRI production	...	...	...	...	...	...	...	...
3. Crude steel production	2 567*	2 945*	2 180	1 941	2 491	2 724	2 457	2 577
4. CC ratio (%)	0*	9*	21	23	23	25	24	35
Iron and Steel Scrap Data								
I. CONSUMPTION (total)	...	...	1 742	1 195	1 402	1 574	1 319	1 477
1. Iron and steel works	...	...	1 742	1 195	1 402	1 574	1 319	1 477
2. Steel foundries	...	...	...	...	...	...	...	...
3. Iron foundries	...	...	...	...	...	...	...	...
4. Re-rolling	...	...	...	...	...	...	...	...
5. Independent electric ferro-alloy	...	...	...	...	...	...	...	...
II. OWN RESOURCES (total)	...	...	119	...	...	123	102	114
1. Circulating scrap	...	...	119	...	...	123	102	114
(a) Iron and steel works	...	...	119	...	...	123	102	114
(b) Steel foundries (independent)	...	...	...	...	...	...	...	...
(c) Iron foundries	...	...	...	...	...	...	...	...
2. Capital scrap (demolition, etc.)	...	...	...	...	...	...	...	...
III. EXTERNAL RECEIPTS (total)	...	...	481	...	...	574	552	394
1. Iron and steel works	...	...	464	...	...	569	544	387
(a) Trade	...	...	...	...	...	...	...	...
(b) Direct purchase	...	...	...	...	...	...	...	...
(c) Imports (not included in (a) or (b))	...	...	...	...	...	...	...	...
2. Steel foundries (independent)	...	...	...	...	...	...	...	...
3. Iron foundries	...	...	...	...	...	...	...	...
4. Independent electric ferro-alloy	...	...	17	...	...	5	8	8
IV. OUTSIDE SALES (-) (total)	...	...	...	...	...	...	...	...
1. Iron and steel works	...	...	...	...	...	...	...	...
2. Steel foundries (independent)	...	...	...	...	...	...	...	...
3. Iron foundries	...	...	...	...	...	...	...	...
V. STOCK VARIATION (total)	...	...	...	...	...	...	...	...
1. Iron and steel works	...	...	...	...	...	...	...	...
2. Steel foundries (independent)	...	...	...	...	...	...	...	...
VI. FOREIGN TRADE (balance)	...	...	1	173	9	11	27	...
1. Imports (-)	...	...	-1	0	0	-1	-17	...
2. Exports (+)	...	...	2	173	9	12	44	...
A. TOTAL NATIONAL RESOURCES (utilized)(II + III + VI)	...	...	601	...	...	708	681	...
B. DOMESTIC SCRAP (total) (A - II.1)(collection utilized)	...	...	...	...	...	...	...	...
C. DOMESTIC SCRAP MARKETED (B - II.2) or (A - II)	...	...	482	...	...	585	579	...
D. EXTERNAL PURCHASE REQUIREMENTS (GROSS)(I - II) (all consumers)	...	...	1 623	...	...	1 451	1 217	1 363
E. EXTERNAL PURCHASE REQUIREMENTS (NET) (I - II + IV) (all consumers)	...	...	...	...	...	...	...	...
F. SURPLUS (+) or DEFICIT (-) (balance) (C - E) or (VI + V)	...	...	...	...	...	...	...	...

Sources: Government replies to UN/ECE questionnaires; * International Iron and Steel Institute.

CANADA

(Thousands of tonnes)

	1980	1985	1990	1993	1994	1995	1996
Basic Iron and Steel Data							
1. Pig iron production	10 892	9 665	7 346	8 633	8 106	8 464	8 638
2. DRI production	...	740*	731*	758*	770*	1000*	1417*
3. Crude steel production	15 902	14 637	12 184	14 296	13 800	14 313	14 637
4. CC ratio (%)	26*	44*	77	92	97	97	98
Iron and Steel Scrap Data							
I. CONSUMPTION (total)	...	7 585	6 554	7 395	7 303	7 351	7 154
1. Iron and steel works	...	7 034	6 554	7 395	7 303	7 351	7 154
2. Steel foundries	...	...	...	...	...	...	...
3. Iron foundries	...	551	...	...	...	...	...
4. Re-rolling	...	...	...	...	...	...	...
5. Independent electric ferro-alloy	...	...	...	...	...	...	...
II. OWN RESOURCES (total)	...	...	2 152	2 428	2 157	2 049	2 075
1. Circulating scrap	...	...	...	...	...	...	...
(a) Iron and steel works	...	...	...	...	...	...	...
(b) Steel foundries (independent)	...	...	...	...	...	...	...
(c) Iron foundries	...	...	...	...	...	...	...
2. Capital scrap (demolition, etc.)	...	...	...	...	...	...	...
III. EXTERNAL RECEIPTS (tota	...	...	4 402	4 967	5 146	5 302	5 079
1. Iron and steel works	...	...	4 402	4 967	5 146	5 302	5 079
(a) Trade	...	...	...	...	...	...	...
(b) Direct purchase	...	...	...	...	...	...	...
(c) Imports (not included in (a) or (b	...	...	1 002	...	...	1 634	1 265
2. Steel foundries (independent)	...	...	...	...	...	...	...
3. Iron foundries	...	...	...	...	...	...	...
4. Independent electric ferro-alloy	...	...	...	...	...	...	...
IV. OUTSIDE SALES (-) (total)	...	...	...	...	...	...	...
1. Iron and steel works	...	...	...	...	...	...	...
2. Steel foundries (independent)	...	...	...	...	...	...	...
3. Iron foundries	...	...	...	...	...	...	...
V. STOCK VARIATION (total)	...	...	...	...	...	...	...
1. Iron and steel works	...	...	...	...	...	...	...
2. Steel foundries (independent)	...	...	...	...	...	...	...
VI. FOREIGN TRADE (balance)	...	...	298	- 166	- 104	487	801
1. Imports (-)	...	...	-1 002	-1 459	-1 739	-1 634	-1 265
2. Exports (+)	...	...	1 301	1 293	1 635	2 121	2 066
A. TOTAL NATIONAL RESOURCES (utilized)(II + III + VI)	...	...	6 852	7229	7 199	7 838	7 955
B. DOMESTIC SCRAP (total) (A - II.1)(collection utilized)	...	...	...	...	...	...	...
C. DOMESTIC SCRAP MARKETED (B - II.2) or (A - II)	...	...	4 700	4 801	5 042	5 789	5 880
D. EXTERNAL PURCHASE REQUIREMENTS (GROSS)(I - II) (all consumers)	...	...	4 402	4 967	5 146	5 302	5 079
E. EXTERNAL PURCHASE REQUIREMENTS (NET) (I - II + IV) (all consumers)	...	...	...	...	...	...	...
F. SURPLUS (+) or DEFICIT (-) (balance) (C - E) or (VI + V)	...	...	...	...	...	...	...

Sources: Government replies to UN/ECE questionnaires; * International Iron and Steel Institute.

Notes: 1980 and 1985 pig iron data also include blast furnace ferro-alloys.

CZECH REPUBLIC

(Thousands of tonnes)

	1990	1991	1992	1993	1994	1995	1996	1997 Forecast
Basic Iron and Steel Data								
1. Pig iron production	6 106	5 317	5 082	4 655	5 268	5 274	4 898	4 900
2. DRI production	-	...	...	...	...	-	-	-
3. Crude steel production	10 098	7 964	7 286	6 769	7 085	7 184	6 509	6 500
4. CC ratio (%)	5	8	10	14	23	32	48	...
Iron and Steel Scrap Data								
I. CONSUMPTION (total)	5 082	...	3 421	3 299	3 244	3 352	3 038	3 100
1. Iron and steel works	5 082	...	3 421	3 299	3 244	3 352	3 038	3 100
2. Steel foundries	...	...	...	...	...	...	...	...
3. Iron foundries	...	...	...	...	...	...	...	...
4. Re-rolling	...	...	...	...	...	...	...	...
5. Independent electric ferro-alloy	...	...	...	...	...	...	...	...
II. OWN RESOURCES (total)	3 068	...	...	...	...	1 671	1 550	1 400
1. Circulating scrap	...	...	...	...	...	...	...	...
(a) Iron and steel works	...	...	...	...	...	...	...	...
(b) Steel foundries (independent)	...	...	...	...	...	...	...	...
(c) Iron foundries	...	...	...	...	...	...	...	...
2. Capital scrap (demolition, etc.)	...	...	...	...	...	...	...	...
III. EXTERNAL RECEIPTS (tota	...	...	...	...	...	...	...	...
1. Iron and steel works	...	...	...	...	...	...	...	...
(a) Trade	...	...	...	...	...	...	...	...
(b) Direct purchase	...	...	...	...	...	...	...	...
(c) Imports (not included in (a) or (b	...	...	...	...	...	...	...	...
2. Steel foundries (independent)	...	...	...	...	...	...	...	...
3. Iron foundries	...	...	...	...	...	...	...	...
4. Independent electric ferro-alloy	...	...	...	...	...	...	...	...
IV. OUTSIDE SALES (-) (total)	...	...	...	...	...	...	...	...
1. Iron and steel works	...	...	...	...	...	...	...	...
2. Steel foundries (independent)	...	...	...	...	...	...	...	...
3. Iron foundries	...	...	...	...	...	...	...	...
V. STOCK VARIATION (total)	...	...	...	...	...	...	...	...
1. Iron and steel works	...	...	...	...	...	...	...	...
2. Steel foundries (independent)	...	...	...	...	...	...	...	...
VI. FOREIGN TRADE (balance)	...	...	...	523	828	700	821	...
1. Imports (-)	...	...	...	- 223	- 158	-89	-84	...
2. Exports (+)	...	...	...	746	986	789	905	...
A. TOTAL NATIONAL RESOURCES (utilized)(II + III + VI)	...	...	...	...	...	...	...	...
B. DOMESTIC SCRAP (total) (A - II.1)(collection utilized)	...	...	...	...	...	...	...	...
C. DOMESTIC SCRAP MARKETED (B - II.2) or (A - II)	...	...	...	...	...	...	...	...
D. EXTERNAL PURCHASE REQUIREMENTS (GROSS)(I - II) (all consumers)	2014	...	...	...	...	1681	1 489	1 700
E. EXTERNAL PURCHASE REQUIREMENTS (NET) (I - II + IV) (all consumers)	...	...	...	...	...	...	...	...
F. SURPLUS (+) or DEFICIT (-) (balance) (C - E) or (VI + V)	...	...	...	...	...	...	...	...

Source: Government replies to UN/ECE questionnaires.

EUROPEAN UNION

(Thousands of tonnes)

	1990	1991	1992	1993	1994	1995	1996
Basic Iron and Steel Data							
1. Pig iron production	91 777	...	...	...	...	97 279	91 652
2. DRI production	...	...	...	...	...	...	...
3. Crude steel production	136 853	...	...	...	...	155 809	146 988
4. CC ratio (%)	90	...	...	...	...	93	94
Iron and Steel Scrap Data							
I. CONSUMPTION (total)	63 929	...	...	...	...	77 285	...
1. Iron and steel works	63 929	...	...	...	...	77 285	...
2. Steel foundries	...	...	...	...	...	...	...
3. Iron foundries	...	...	...	...	...	...	...
4. Re-rolling	...	...	...	...	...	...	...
5. Independent electric ferro-alloy	...	...	...	...	...	...	...
II. OWN RESOURCES (total)	14 348	...	...	...	...	15 262	...
1. Circulating scrap	...	...	...	...	...	...	...
(a) Iron and steel works	...	...	...	...	...	...	...
(b) Steel foundries (independent)	...	...	...	...	...	...	...
(c) Iron foundries	...	...	...	...	...	...	...
2. Capital scrap (demolition, etc.)	...	...	...	...	...	...	...
III. EXTERNAL RECEIPTS (total)	64 278	...	...	...	...	82 028	...
1. Iron and steel works	64 278	...	...	...	...	82 028	...
(a) Trade	...	...	...	...	...	...	...
(b) Direct purchase	...	...	...	...	...	...	...
(c) Imports (not included in (a) or (b))	16 435	...	...	...	...	23 303	...
2. Steel foundries (independent)	...	...	...	...	...	...	...
3. Iron foundries	...	...	...	...	...	...	...
4. Independent electric ferro-alloy	...	...	...	...	...	...	...
IV. OUTSIDE SALES (-) (total)	...	...	...	...	...	...	...
1. Iron and steel works	...	...	...	...	...	...	...
2. Steel foundries (independent)	...	...	...	...	...	...	...
3. Iron foundries	...	...	...	...	...	...	...
V. STOCK VARIATION (total)	...	...	...	...	...	...	...
1. Iron and steel works	-286	...	...	...	...	8 304	...
2. Steel foundries (independent)	...	...	...	...	...	...	...
VI. FOREIGN TRADE (balance)	-395	...	...	...	...	1882	1 644
1. Imports (-)	-3 646	...	...	...	...	-5178	-3 389
2. Exports (+)	3 251	...	...	...	...	7060	5 033
A. TOTAL NATIONAL RESOURCES (utilized)(II + III + VI)	78 231	...	...	...	...	99 172	...
B. DOMESTIC SCRAP (total) (A - II.1)(collection utilized)	...	...	...	...	...	...	...
C. DOMESTIC SCRAP MARKETED (B - II.2) or (A - II)	63 883	...	...	...	...	83 910	...
D. EXTERNAL PURCHASE REQUIREMENTS (GROSS)(I - II) (all consumers)	49 581	...	...	...	...	62 023	...
E. EXTERNAL PURCHASE REQUIREMENTS (NET) (I - II + IV) (all consumers)	...	...	...	...	...	...	...
F. SURPLUS (+) or DEFICIT (-) (balance) (C - E) or (VI + V)	...	...	...	...	...	...	...

Source: Government replies to UN/ECE questionnaires.

Notes: 1990 data for EUR12, 1996 foreign trade for 9 months.

FINLAND

(Thousands of tonnes)

	1980	1985	1990	1993	1994	1995	1996	1997 Forecast
Basic Iron and Steel Data								
1. Pig iron production	2019*	1901*	2 381	2 535	2 597	2 242	2 457	2 744
2. DRI production	...	...	...	...	...	...	...	
3. Crude steel production	2508*	2518*	2 846	3 256	3 420	3 166	3 296	3 649
4. CC ratio (%)	90*	94*	100	99	99	100	100	100
Iron and Steel Scrap Data								
I. CONSUMPTION (total)	...	...	...	...	...	1 134	1 128	...
1. Iron and Steel Works	...	...	...	...	...	1 118	1 115	...
2. Steel foundries	...	...	...	...	...	16	13	13
3. Iron foundries	...	...	-	...	...	-	-	-
4. Re-rolling	...	...	-	...	...	-	-	-
5. Independent electric ferro-alloy	...	...	-	...	...	-	-	-
II. OWN RESOURCES (total)	...	...	429	472	439	430	434	...
1. Circulating scrap	...	...	...	...	...	426	429	...
(a) Iron and steel works	...	...	...	...	...	421	425	...
(b) Steel foundries (independent)	...	...	...	...	...	5	4	4
(c) Iron foundries	...	...	-	...	...	-	-	-
2. Capital scrap (demolition, etc.)	...	...	...	...	...	4	5	...
III. EXTERNAL RECEIPTS (tota	...	...	454	429	558	460	546	458
1. Iron and steel works	...	...	392	382	500	460	546	458
(a) Trade	...	...	392	382	500	324	333	346
(b) Direct purchase	...	...	...	...	...	12	15	12
(c) Imports (not included in (a) or (b	...	...	...	...	...	124	198	100
2. Steel foundries (independent)	...	...	62	47	58	-	-	-
3. Iron foundries	...	...	...	...	...	-	-	-
4. Independent electric ferro-alloy	...	...	...	...	...	-	-	-
IV. OUTSIDE SALES (-) (total)	...	...	...	...	...	18	20	21
1. Iron and steel works	...	...	...	...	...	18	20	21
2. Steel foundries (independent)	...	...	...	...	...	-	-	-
3. Iron foundries	...	...	...	...	...	-	-	-
V. STOCK VARIATION (total)	...	...	...	...	...	-262	-168	...
1. Iron and steel works	...	...	...	...	...	23	-12	7
2. Steel foundries (independent)	...	...	...	...	...	-	-	-
VI. FOREIGN TRADE (balance)	...	...	-44	-93	-230	-117	6	...
1. Imports (-)	...	...	-46	-140	-338	-297	-223	...
2. Exports (+)	...	...	2	46	108	180	228	...
A. TOTAL NATIONAL RESOURCES								
(utilized)(II + III + VI)	...	...	839	808	767	773	986	...
B. DOMESTIC SCRAP (total)								
(A - II.1)(collection utilized)	...	...	...	...	...	347	557	...
C. DOMESTIC SCRAP MARKETE								
(B - II.2) or (A - II)	...	...	410	336	328	343	552	...
D. EXTERNAL PURCHASE REQUIREMENTS								
(GROSS)(I - II) (all consumers)	...	...	...	...	...	704	694	...
E. EXTERNAL PURCHASE REQUIREMENTS								
(NET) (I - II + IV) (all consumers	...	...	...	...	...	722	714	...
F. SURPLUS (+) or DEFICIT (-) (balance)								
(C - E) or (VI + V)	...	...	...	...	...	-379	-162	...

Sources: Government replies to UN/ECE questionnaires; * International Iron and Steel Institute.

FRANCE

(Thousands of tonnes)

	1980	1985	1990	1993	1994	1995	1996
Basic Iron and Steel Data							
1. Pig iron production	19 157	15 405	14 415	12 664	13 293	13 154	12 402
2. DRI production	...	...	...	...	...	...	...
3. Crude Steel production	23 176	18 808	18 763	17 106	18 031	17 872	17 633
4. CC Ratio (%)	41	81	96	95	95	95	95
Iron and Steel Scrap Data							
I. CONSUMPTION (total)	10 368	8 509	9 128	8 797	9 728	10 023	9 826
1. Iron and Steel Works	7 901	6 478	7 204	6 925	7 708	7 838	7 769
2. Steel foundries	309	208	250	191	193	197	190
3. Iron foundries	2 044	1 695	1 583	1 681	1 827	1 899	1 774
4. Re-rolling	...	...	65	...	...	65	66
5. Independent Electric Ferro-Alloy	...	...	26	...	...	24	27
II. OWN RESOURCES (total)	5 614	3 988	3 233	2 959	3 145	3 165	2 926
1. Circulating scrap	4 914	3 593	2 943	2 580	2 855	2 860	2 642
(a) Iron and steel works	3 935	2 846	2 106	1 886	2 094	2 073	1 876
(b) Steel foundries (independent)	144	94	122	82	80	78	83
(c) Iron foundries	835	653	715	612	681	709	683
2. Capital scrap (demolition, etc.)	700	394	290	379	290	305	284
III. EXTERNAL RECEIPTS (total)	4 778	4 628	6 552	6 753	7 440	7 545	7 483
1. Iron and steel works	3 404	3 472	5 529	5 554	6 155	6 212	6 258
(a) Trade	2 963	3 207	4 700	4 150	4 476	4 526	4 648
(b) Direct purchase	367	262	810	1 404	1 679	1 684	1 588
(c) Imports (not included in (a) or (b))	74	3	19	-	-	2	22
2. Steel foundries (independent)	165	114	128	109	113	119	107
3. Iron foundries	1 209	1 042	868	1 069	1 146	1 190	1 091
4. Independent Electric Ferro-Alloy	...	...	27	21	26	24	27
IV. OUTSIDE SALES (-) (total)	231	208	...	-	20	20	28
1. Iron and steel works	230	208	-	-	20	20	28
2. Steel foundries (independent)	-	-	...	-	-	...	...
3. Iron foundries	-	-	...	-	-	...	...
V. STOCK VARIATION (total)	- 207	- 101	729	894	837	667	555
1. Iron and steel works	- 92	26	26	69	28	- 5	-71
2. Steel foundries (independent)	-	-	...	-	-	...	...
VI. FOREIGN TRADE (balance)	2 856	3 481	2 490	2 513	2 183	2 055	2 467
1. Imports (-)	- 499	- 496	- 980	-1 096	-1 378	-1 627	-1 273
2. Exports (+)	3 355	3 977	3 470	3 609	3 561	3 682	3 740
A. TOTAL NATIONAL RESOURCES (utilized)(II + III + VI)	13 248	12 097	12 275	12 225	12 768	12 765	12 876
B. DOMESTIC SCRAP (total) (A - II.1)(collection utilized)	8 334	8 504	9 332	9 645	9 913	9 905	10 234
C. DOMESTIC SCRAP MARKETED (B - II.2) or (A - II)	7 634	8 109	9 042	9 266	9 623	9 600	9 950
D. EXTERNAL PURCHASE REQUIREMENTS (GROSS)(I - II) (all consumers)	4 754	4 521	5 895	5 838	6 583	6 858	6 900
E. EXTERNAL PURCHASE REQUIREMENTS (NET) (I - II + IV) (all consumers)	4 985	4 729	5 895	5 838	6 603	6 878	6 928
F. SURPLUS (+) or DEFICIT (-) (balance) (C - E) or (VI + V)	2 649	3 380	3 147	3 428	3 020	2 722	3 022

Source: Government replies to UN/ECE questionnaires.

Note: Stock variation calculated as follows: V = II + III - I - IV.

GERMANY

(Thousands of tonnes)

	1980	1985	1990	1993	1994	1995	1996
Basic Iron and Steel Data							
1. Pig iron production	33 873	31 531	30 097	26 970	29 923	30 012	27 722
2. DRI production*	...	100	310	182	281	410	370
3. Crude steel production	43 838	40 497	38 434	37 625	40 837	42 051	39 793
4. CC ratio (%)	46	80	91	94	96	95	96
Iron and Steel Scrap Data							
I. CONSUMPTION (total)	20 093	18 258	16 672	17 945	18 711	20 220	...
1. Iron and steel works	15 657	14 117	12 201	14 276	14 616	15 725	15 749
2. Steel foundries	282	363	334	272	280	284	...
3. Iron foundries	4 154	3 778	4 137	3 397	3 815	4 211	...
4. Re-rolling	-	-	-	-	-	-	-
5. Independent electric ferro-alloy	-	-	-	-	-	-	-
II. OWN RESOURCES (total)	9 430	7 845	6 517	6 234	6 509	6 600	...
1. Circulating scrap	8 039	7 047	6 388	5 950	6 280	6 434	...
(a) Iron and steel works	6 064	5 538	4 613	4 490	4 665	4 718	4 490
(b) Steel foundries (independent)	215	166	165	134	180	126	...
(c) Iron foundries	1 760	1 343	1 610	1 326	1 435	1 590	...
2. Capital scrap (demolition, etc.)	1 391	798	129	284	229	166	182
III. EXTERNAL RECEIPTS (total)	10 140	10 382	10 347	12 037	12 930	14 129	...
1. Iron and steel works	7 763	8 267	7 921	10 046	10 478	11 467	11 600
(a) Trade	7 661	8 075	7 442	9 511	9 883	10 886	11 111
(b) Direct purchase	78	192	366	351	463	434	361
(c) Imports (not included in (a) or (b))	24	-	113	184	132	147	128
2. Steel foundries (independent)	247	197	178	139	177	162	...
3. Iron foundries	2 130	1 888	2 248	1 852	2 275	2 500	...
4. Independent electric ferro-alloy	-	-	-	-	-	-	-
IV. OUTSIDE SALES (-) (total)	230	490	520	559	666	609	566
1. Iron and steel works	230	490	520	559	666	609	566
2. Steel foundries (independent)	-	-	-	-	-	-	-
3. Iron foundries	-	-	-	-	-	-	...
V. STOCK VARIATION (total)	- 753	- 521	- 328	- 233	62	- 100	...
1. Iron and steel works	-1 020	- 141	- 58	15	90	19	- 42
2. Steel foundries (independent)	217	11	9	1	77	4	...
VI. FOREIGN TRADE (balance)	1 572	1 795	4 002	7 610	7 702	6 768	5 349
1. Imports (-)	-1 506	-1 614	-1 041	- 910	-1 017	-1 197	-1 335
2. Exports (+)	3 078	3 409	5 043	8 520	8 719	7 965	6 684
A. TOTAL NATIONAL RESOURCES (utilized)(II + III + VI)	21 142	20 022	20 866	25 881	27 141	27 497	...
B. DOMESTIC SCRAP (total) (A - II.1)(collection utilized)	13 103	12 975	14 478	19 931	20 861	21 063	...
C. DOMESTIC SCRAP MARKETED (B - II.2) or (A - II)	11 712	12 177	14 349	19 647	20 632	20 897	...
D. EXTERNAL PURCHASE REQUIREMENTS (GROSS)(I - II) (all consumers)	10 663	10 413	10 155	11 711	12 202	13 620	...
E. EXTERNAL PURCHASE REQUIREMENTS (NET) (I - II + IV) (all consumers)	10 893	10 903	10 675	12 270	12 868	14 229	...
F. SURPLUS (+) or DEFICIT (-) (balance) (C - E) or (VI + V)	819	1 274	3 674	7 377	7 764	6 668	...

Sources: Government replies to UN/ECE questionnaires; * International Iron and Steel Institute.

Notes: Stock variation calculated as follows: V = II + III - I - IV.

1980 and 1985 pig iron data also include blast furnace ferro-alloys.

Data for 1980 to 1990 are for the former West Germany, as communicated.

GREECE

(Thousands of tonnes)

	1980	1985	1990	1993	1994	1995	1996
Basic Iron and Steel Data							
1. Pig iron production	303	-	-	-	-	-	-
2. DRI production	...	...	...	...	...	...	...
3. Crude steel production	870	955	999	980	848	939	847
4. CC ratio (%)	...	...	100	100	100	100	100
Iron and Steel Scrap Data							
I. CONSUMPTION (total)	...	...	1 095	1 094	973	1 070	964
1. Iron and steel works	...	...	1 095	1 094	973	1 070	964
2. Steel foundries	...	...	...	...	...	...	...
3. Iron foundries	...	...	...	...	...	...	...
4. Re-rolling	...	...	...	...	...	...	...
5. Independent electric ferro-alloy	...	...	...	...	...	...	...
II. OWN RESOURCES (total)	...	...	106	86	98	103	95
1. Circulating scrap	...	...	106	86	93	94	95
(a) Iron and steel works	...	...	106	86	93	94	95
(b) Steel foundries (independent)	...	...	...	...	...	...	...
(c) Iron foundries	...	...	...	...	...	...	...
2. Capital scrap (demolition, etc.)	...	...	-	-	4	10	0
III. EXTERNAL RECEIPTS (total)	...	...	993	1 001	910	961	939
1. Iron and steel works	...	...	993	1 001	910	961	939
(a) Trade	...	...	320	510	423	485	422
(b) Direct purchase	...	...	13	12	14	-	-
(c) Imports (not included in (a) or (b))	...	...	660	479	473	476	517
2. Steel foundries (independent)	...	...	...	...	...	...	...
3. Iron foundries	...	...	...	...	...	...	...
4. Independent Electric Ferro-Alloy	...	...	...	...	...	...	...
IV. OUTSIDE SALES (-) (total)	...	...	24	15	20	4	4
1. Iron and steel works	...	...	24	15	20	4	4
2. Steel foundries (independent)	...	...	...	...	...	...	...
3. Iron foundries	...	...	...	...	...	...	...
V. STOCK VARIATION (total)	...	...	- 19	- 22	15	- 10	67
1. Iron and steel works	...	...	- 19	- 22	15	- 10	67
2. Steel foundries (independent)	...	...	...	...	...	...	...
VI. FOREIGN TRADE (balance)	...	...	-1 006	- 352	- 305	- 438	...
1. Imports (-)	...	...	-1 752	- 366	- 347	- 459	...
2. Exports (+)	...	...	746	14	42	21	...
A. TOTAL NATIONAL RESOURCES (utilized)(II + III + VI)	...	...	93	735	703	626	...
B. DOMESTIC SCRAP (total) (A - II.1)(collection utilized)	...	...	- 13	649	609	532	...
C. DOMESTIC SCRAP MARKETED (B - II.2) or (A - II)	...	...	- 13	649	605	523	...
D. EXTERNAL PURCHASE REQUIREMENTS (GROSS)(I - II) (all consumers)	...	...	988	1 008	875	967	869
E. EXTERNAL PURCHASE REQUIREMENTS (NET) (I - II + IV) (all consumers)	...	...	1 012	1 023	895	971	873
F. SURPLUS (+) or DEFICIT (-) (balance) (C - E) or (VI + V)	...	...	-1 025	- 374	- 290	- 448	...

Source: Government replies to UN/ECE questionnaires.

Note: Stock variation calculated as follows: V = II + III - I - IV.

HUNGARY

(Thousands of tonnes)

	1980	1985	1990	1993	1994	1995	1996	1997 Forecast
Basic Iron and Steel Data								
1. Pig iron production	2 214	2 102	1 893	1 407	1 595	1 515	1 496	1037
2. DRI production	-	-	-	-	-	-	-	-
3. Crude steel production	3 767	3 590	2 823	1 748	1 924	1 861	1 872	1720
4. CC ratio (%)	36	47	77	87	89	91	96	96
Iron and Steel Scrap Data								
I. CONSUMPTION (total)	...	2 050	1 533	599	619	770	755	1 004
1. Iron and steel works	...	2 050	1 365	535	559	712	755	1 004
2. Steel foundries	...	...	29	13	14	14	...	...
3. Iron foundries	...	...	139	51	46	44	...	...
4. Re-rolling	...	...	-	-	-	-	-	-
5. Independent electric ferro-alloy	...	...	-	-	-	-		
II. OWN RESOURCES (total)	...	1 066	779	366	347	256	242	231
1. Circulating scrap	...	917	39	17	17	206	215	192
(a) Iron and steel works	...	917	...	...	...	189	215	192
(b) Steel foundries (independent)	...	...	20	8	9	9	...	...
(c) Iron foundries	...	...	20	8	9	9	...	...
2. Capital scrap (demolition, etc.)	...	149	...	...	...	50	27	39
III. EXTERNAL RECEIPTS (tota	...	1 084	69	26	24	520	532	797
1. Iron and steel works	...	1 084	...	...	...	497	532	797
(a) Trade	...	923	...	...	...	392	423	522
(b) Direct purchase	...	145	...	...	...	105	109	275
(c) Imports (not included in (a) or (b	...	16	-	-	-	-		-
2. Steel foundries (independent)	...	...	10	5	6	5	...	...
3. Iron foundries	...	...	60	21	19	18	...	...
4. Independent electric ferro-alloy	...	...	-	-	-	-	-	-
IV. OUTSIDE SALES (-) (total)	...	76						
1. Iron and steel works	...	76	-	-	-	-	-	
2. Steel foundries (independent)	...	...	-	-	-	-	-	
3. Iron foundries	...	...	-	-	-	-	-	-
V. STOCK VARIATION (total)	...	24	- 685	- 207	- 247	6	19	...
1. Iron and steel works	...	24	-	-	-	23	22	25
2. Steel foundries (independent)	...	...	...	...	...	...	...	...
VI. FOREIGN TRADE (balance)	...	45	- 411	549	660	748	601	...
1. Imports (-)	...	...	- 605	- 4	- 6	-9	-12	...
2. Exports (+)	...	45	194	553	666	757	613	...
A. TOTAL NATIONAL RESOURCES (utilized)(II + III + VI)	...	2 195	437	941	1 031	1 524	1 375	1 028
B. DOMESTIC SCRAP (total) (A - II.1)(collection utilized)	...	1 278	398	924	1 014	1 318	1 160	836
C. DOMESTIC SCRAP MARKED (B - II.2) or (A - II)	...	1 129	- 342	575	684	1 268	1 133	
D. EXTERNAL PURCHASE REQUIREMENTS (GROSS)(I - II) (all consumers)	...	984	754	233	272	514	513	773
E. EXTERNAL PURCHASE REQUIREMENTS (NET) (I - II + IV) (all consumers)	...	1 060	754	233	272	514	513	773
F. SURPLUS (+) or DEFICIT (-) (balance) (C - E) or (VI + V)	...	69	-1 096	342	413	754	620	

Source: Government replies to UN/ECE questionnaires.

Note: Stock variation calculated as follows: V = II + III - I - IV.

IRELAND

(Thousands of tonnes)

	1980*	1985*	1990	1993	1994	1995	1996	1997 Forecast
Basic Iron and Steel Data								
1. Pig iron production	-	-	-	-	-	-	-	...
2. DRI production	...	...	...	...	...	-	-	...
3. Crude steel production	2	203	325	326	326	310	341	...
4. CC ratio (%)	-	100	100	100	100	100	100	...
Iron and Steel Scrap Data								
I. CONSUMPTION (total)	...	...	...	...	...	337	389	...
1. Iron and steel works	...	...	...	...	...	337	389	...
2. Steel foundries	...	...	...	...	...	-	-	-
3. Iron foundries	...	...	...	...	...	-	-	-
4. Re-rolling	...	...	...	...	...	-	-	-
5. Independent electric ferro-alloy	...	...	...	...	...	-	-	-
II. OWN RESOURCES (total)	...	...	...	...	...	9	13	23
1. Circulating scrap	...	...	...	...	...	...	...	...
(a) Iron and steel works	...	...	...	...	...	-	-	...
(b) Steel foundries (independent)	...	...	...	...	...	...	...	...
(c) Iron foundries	...	...	...	...	...	...	...	...
2. Capital scrap (demolition, etc.)	...	...	...	...	...	9	13	23
III. EXTERNAL RECEIPTS (tota	...	...	...	...	...	296	344	381
1. Iron and steel works	...	...	...	...	...	296	344	381
(a) Trade	...	...	...	...	...	...	...	...
(b) Direct purchase	...	...	...	...	...	...	...	...
(c) Imports (not included in (a) or (b	...	...	...	...	...	...	...	...
2. Steel foundries (independent)	...	...	...	...	...	-	-	-
3. Iron foundries	...	...	...	...	...	-	-	-
4. Independent electric ferro-alloy	...	...	...	...	...	-	-	-
IV. OUTSIDE SALES (-) (total)	...	...	...	...	...	...	...	...
1. Iron and steel works	...	...	...	...	...	-	-	-
2. Steel foundries (independent)	...	...	...	...	...	-	-	-
3. Iron foundries	...	...	...	...	...	-	-	-
V. STOCK VARIATION (total)	...	...	...	...	...	-32	-32	...
1. Iron and steel works	...	...	...	...	...	...	...	...
2. Steel foundries (independent)	...	...	...	...	...	-	-	-
VI. FOREIGN TRADE (balance)	...	...	-	...	-191	-176	-178	-213
1. Imports (-)	...	...	-	...	-191	-176	-178	-213
2. Exports (+)	...	...	-	...	...	-	-	-
A. TOTAL NATIONAL RESOURCES (utilized)(II + III + VI)	...	...	...	...	...	129	179	191
B. DOMESTIC SCRAP (total) (A - II.1)(collection utilized)	...	...	...	...	...	129	179	191
C. DOMESTIC SCRAP MARKETE (B - II.2) or (A - II)	...	...	...	...	...	120	166	168
D. EXTERNAL PURCHASE REQUIREMENTS (GROSS)(I - II) (all consumers)	...	...	...	...	...	328	376	...
E. EXTERNAL PURCHASE REQUIREMENTS (NET) (I - II + IV) (all consumers	...	...	...	...	...	328	376	...
F. SURPLUS (+) or DEFICIT (-) (balance) (C - E) or (VI + V)	...	...	...	...	...	-208	-210	...

Sources: Government replies to UN/ECE questionnaires. * International Iron and Steel Institute.

ITALY

(Thousands of tonnes)

	1980	1985	1990	1993	1994	1995	1996
Basic Iron and Steel Data							
1. Pigiron production	12 218	12 114	11 852	11 195	11 186	11 664	10 321
2. DRI production	...	...	-	-	-	-	-
3. Crude steel production	26 501	23 898	25 340	25 720	26 149	27 673	24 196
4. CC ratio (%)	50*	79*	95	97	96	97	97
Iron and Steel Scrap Data							
I. CONSUMPTION (total)	19 207	16 380	16 698	17 434	17 675	18 182	15 879
1. Iron and Steel Works	17 748	15 270	16 436	17 281	17 549	18 072	15 878
2. Steel foundries	157	133	126	87	100	99	...
3. Iron foundries	1 200	810	...	...	...	...	...
4. Re-rolling	...	...	136	66	26	11	1
5. Independent electric ferro-alloy	...	...	...	...	...	...	...
II. OWN RESOURCES (total)	5 920	4 084	2 838	2 344	2 534	2 350	2 034
1. Circulating scrap	5 169	3 346	2 650	2 266	2 431	2 282	2 008
(a) Iron and steel works	4 458	2 981	2 567	2 207	2 370	2 253	2 008
(b) Steel foundries (independent)	71	...	83	59	61	29	...
(c) Iron foundries	640	365	...	...	...	...	...
2. Capital scrap (demolition, etc.)	751	738	188	78	103	68	26
III. EXTERNAL RECEIPTS (tota	13 558	13 016	13 985	14 963	15 319	15 680	13 704
1. Iron and steel works	12 799	12 271	13 932	14 932	15 282	15 609	13 704
(a) Trade	10 585	9 564	7 158	8 409	7 927	8 824	8 607
(b) Direct purchase	451	1 359	955	1 228	1 141	1 006	811
(c) Imports (not included in (a) or (b	1 763	1 348	5 819	5 295	6 214	5 779	4 286
2. Steel foundries (independent)	97	133	53	31	37	71	...
3. Iron foundries	560	445	...	...	...	...	...
4. Independent electric ferro-alloy	...	...	...	...	...	...	...
IV. OUTSIDE SALES (-) (total)	472	445	234	57	54	74	77
1. Iron and steel works	472	445	234	57	54	74	77
2. Steel foundries (independent)	...	...	-	-	-	-	...
3. Iron foundries	...	...	...	...	...	...	...
V. STOCK VARIATION (total)	- 201	275	- 109	- 184	124	- 226	- 218
1. Iron and steel works	- 212	275	- 28	72	275	- 96	- 66
2. Steel foundries (independent)	11	-	- 1	1	...	- 1	1
VI. FOREIGN TRADE (balance)	-7 402	-5 463	-5 788	-5 891	-6 155	-6 304	-4 506
1. Imports (-)	-7 410	-5 473	-5 807	-5 904	-6 165	-6 337	-4 524
2. Exports (+)	8	10	19	13	10	33	18
A. TOTAL NATIONAL RESOURCES							
(utilized)(II + III + VI)	12 076	11 637	11 035	11 416	11 698	11 726	11 232
B. DOMESTIC SCRAP (total)							
(A - II.1)(collection utilized)	6 907	8 291	8 385	9 150	9 267	9 444	9 224
C. DOMESTIC SCRAP MARKETED							
(B - II.2) or (A - II)	6 156	7 553	8 197	9 072	9 164	9 376	9 198
D. EXTERNAL PURCHASE REQUIREMENTS							
(GROSS)(I - II) (all consumers)	13 287	12 296	13 860	15 090	15 141	15 832	13 845
E. EXTERNAL PURCHASE REQUIREMENTS							
(NET) (I - II + IV) (all consumers	13 759	12 741	14 094	15 147	15 195	15 906	13 922
F. SURPLUS (+) or DEFICIT (-) (balance)							
(C - E) or (VI + V)	-7 603	-5 188	-5 897	-6 075	-6 031	-6 530	-4 724

Sources: Government replies to UN/ECE questionnaires; * International Iron and Steel Institute.

Notes: Stock variation calculated as follows: V = II + III - I - IV.

1980 and 1985 pig iron data also include blast furnace ferro-alloys.

JAPAN

(Thousands of tonnes)

	1980	1985	1990	1993	1994	1995	1996
Basic Iron and Steel Data							
1. Pig iron production	87 041	80 569	80 229	73 738	73 776	74 905	74 597
2. DRI production	...	...	-	-	-	-	-
3. Crude steel production	111 395	105 279	110 339	99 623	98 294	101 640	98 801
4. CC ratio (%)	60	91	94	96	96	96	96
Iron and Steel Scrap Data							
I. CONSUMPTION (total)	43 237	43 623	48 204	43 086	43 074	43 753	43 866
1. Iron and steel works	36 238	36 713	40 011	36 409	36 388	37 004	37 062
2. Steel foundries	...	...	...	...	...	...	...
3. Iron foundries	6 207	6 400	7 634	6 293	6 330	6 438	6 498
4. Re-rolling	...	...	559	384	356	311	306
5. Independent electric ferro-alloy	...	...	...	...	...	...	...
II. OWN RESOURCES (total)	15 435	12 015	13 226	12 374	12 478	12 393	12 705
1. Circulating scrap	15 435	12 015	13 226	12 374	12 478	12 393	12 705
(a) Iron and steel works	13 264	9 753	10 225	9 862	9 925	9 776	10 080
(b) Steel foundries (independent)	...	...	...	...	...	...	...
(c) Iron foundries	2 171	2 262	3 001	2 512	2 553	2 617	2 625
2. Capital scrap (demolition, etc.)	...	...	...	...	...	...	...
III. EXTERNAL RECEIPTS (tota	30 401	34 049	37 540	33 757	33 863	34 986	34 971
1. Iron and steel works	25 599	29 476	32 907	29 976	30 086	31 165	31 098
(a) Trade	25 599	26 563	32 111	29 161	29 129	30 202	30 888
(b) Direct purchase	...	...	...	...	...	...	...
(c) Imports (not included in (a) or (b	...	2 913	796	815	957	963	210
2. Steel foundries (independent)	...	...	...	...	...	...	...
3. Iron foundries	4 045	4 138	4 633	3 781	3 777	3 821	3 873
4. Independent electric ferro-alloy	...	...	...	...	...	...	...
IV. OUTSIDE SALES (-) (total)	3 645	2 784	...	...	...	...	...
1. Iron and steel works	3 645	2 784	...	...	...	...	...
2. Steel foundries (independent)	...	...	...	...	...	...	...
3. Iron foundries	...	...	...	...	...	...	...
V. STOCK VARIATION (total)	-1 046	- 343	...	...	...	...	...
1. Iron and steel works	-1 020	- 268	- 145	- 99	- 174	368	- 97
2. Steel foundries (independent)	...	...	...	...	...	...	...
VI. FOREIGN TRADE (balance)	-2 737	-3 088	- 652	265	- 101	- 294	1 618
1. Imports (-)	-2 896	-3 254	-1 048	- 913	-1 068	-1 209	- 375
2. Exports (+)	159	166	396	1 178	967	915	1 993
A. TOTAL NATIONAL RESOURCES (utilized)(II + III + VI)	43 099	42 976	50 114	46 396	46 240	47 085	49 294
B. DOMESTIC SCRAP (total) (A - II.1)(collection utilized)	27 664	30 961	36 888	34 022	33 762	34 692	36 589
C. DOMESTIC SCRAP MARKETED (B - II.2) or (A - II)	27 664	30 961	36 888	34 022	33 762	34 692	36 589
D. EXTERNAL PURCHASE REQUIREMENTS (GROSS)(I - II) (all consumers)	27 802	31 608	34 978	30 712	30 596	31 360	31 161
E. EXTERNAL PURCHASE REQUIREMENTS (NET) (I - II + IV) (all consumers)	31 447	34 392	...	...	...	...	...
F. SURPLUS (+) or DEFICIT (-) (balance) (C - E) or (VI + V)	-3 783	-3 431	...	...	...	...	...

Source: Government replies to UN/ECE questionnaires.

Notes: Stock variation calculated as follows: V = II + III - I - IV.

1980 and 1985 pig iron data also include blast furnace ferro-alloys. Item III 1(a) includes item III 1(b).

LATVIA

(Thousands of tonnes)

	1985	1990	1992	1993	1994	1995	1996
Basic Iron and Steel Data							
1. Pig iron production	...	-	-	-	-	-	-
2. DRI production	...	-	-	-	-	-	-
3. Crude steel production	...	557	246	300	332	279	293
4. CC ratio (%)	...	97	97	99	99	99	99
Iron and Steel Scrap Data							
I. CONSUMPTION (total)	...	457	243	310	350	315	334
1. Iron and steel works	...	457	243	310	350	315	334
2. Steel foundries	...	-	-	-	-	...	...
3. Iron foundries	...	-	-	-	-	...	...
4. Re-rolling	...	-	-	-	-	...	...
5. Independent electric ferro-alloy	...	-	-	-	-	...	...
II. OWN RESOURCES (total)	...	273	...	...	...	43	46
1. Circulating scrap	...	207	...	...	...	38	39
(a) Iron and steel works	...	207	...	...	...	38	39
(b) Steel foundries (independent)	...	-	-	-	-	...	...
(c) Iron foundries	...	-	-	-	-	...	...
2. Capital scrap (demolition, etc.)	...	66	...	...	...	5	7
III. EXTERNAL RECEIPTS (total)	...	365	165	280	359	...	...
1. Iron and steel works	...	365	165	280	359	...	...
(a) Trade	...	...	...	...	...	...	...
(b) Direct purchase	...	...	...	...	...	...	...
(c) Imports (not included in (a) or (b))	...	...	...	...	...	...	...
2. Steel foundries (independent)	...	-	-	-	-	...	...
3. Iron foundries	...	-	-	-	-	...	...
4. Independent electric ferro-alloy	...	-	-	-	-	...	...
IV. OUTSIDE SALES (-) (total)	...	188	162	-	-	...	...
1. Iron and steel works	...	188	162	-	-	...	...
2. Steel foundries (independent)	...	-	-	-	-	...	...
3. Iron foundries	...	-	-	-	-	...	...
V. STOCK VARIATION (total)	...	-7	...	...	...	...	...
1. Iron and steel works	...	...	...	...	...	...	...
2. Steel foundries (independent)	...	-	-	-	-	...	...
VI. FOREIGN TRADE (balance)	...	...	...	-20	15	1	6
1. Imports (-)	...	...	...	-27	-4	-	-
2. Exports (+)	...	...	...	7	19	1	6
A. TOTAL NATIONAL RESOURCES (utilized)(II + III + VI)	...	...	...	...	...	...	...
B. DOMESTIC SCRAP (total) (A - II.1)(collection utilized)	...	...	...	...	...	...	...
C. DOMESTIC SCRAP MARKETED (B - II.2) or (A - II)	...	...	...	...	...	...	...
D. EXTERNAL PURCHASE REQUIREMENTS (GROSS)(I - II) (all consumers)	...	184	...	...	...	272	288
E. EXTERNAL PURCHASE REQUIREMENTS (NET) (I - II + IV) (all consumers)	...	372	...	...	...	...	...
F. SURPLUS (+) or DEFICIT (-) (balance) (C - E) or (VI + V)	...	...	...	...	...	...	...

Source: Government replies to UN/ECE questionnaires.

Note: Stock variation calculated as follows: V = II + III - I - IV.

LUXEMBOURG

(Thousands of tonnes)

	1980	1985	1990	1993	1994	1995	1996
Basic Iron and Steel Data							
1. Pig iron production	3 568	2 754	2 645	2 412	1 927	1 028	829
2. DRI production	...	...	-	-	-	-	-
3. Crude steel production	4 619	3 945	3 560	3 293	3 073	2 613	2 502
4. CC ratio (%)	-	28	34	38	30	46	53
Iron and Steel Scrap Data							
I. CONSUMPTION (total)	1 577	1 625	1 387	1 278	1 510	1 958	2 011
1. Iron and steel works	1 545	1 571	1 373	1 278	1 510	1 958	2 011
2. Steel foundries	...	...	...	...	...	...	...
3. Iron foundries	32	54	14	...	...	...	...
4. Re-rolling	...	...	...	...	...	...	...
5. Independent electric ferro-alloy	...	...	...	...	...	...	...
II. OWN RESOURCES (total)	1 111	746	667	579	601	544	470
1. Circulating scrap	843	536	487	465	492	465	390
(a) Iron and steel works	841	530	487	465	492	460	386
(b) Steel foundries (independent)	...	...	...	...	...	5	4
(c) Iron foundries	2	6	-	-	-	-	-
2. Capital scrap (demolition, etc.)	268	210	180	114	109	79	80
III. EXTERNAL RECEIPTS (total)	594	903	726	789	1 063	1 491	1 618
1. Iron and steel works	564	868	711	789	1 063	1 491	1 618
(a) Trade	564	818	-	-	-	-	-
(b) Direct purchase	-	50	82	112	96	89	63
(c) Imports (not included in (a) or (b))	...	...	629	677	967	1 402	1 555
2. Steel foundries (independent)	...	...	...	...	...	...	...
3. Iron foundries	30	35	15	...	...	...	...
4. Independent electric ferro-alloy	...	...	...	...	...	...	...
IV. OUTSIDE SALES (-) (total)	133	52	23	52	89	66	67
1. Iron and steel works	133	43	22	52	89	66	67
2. Steel foundries (independent)	...	...	...	...	...	...	...
3. Iron foundries	...	9	1	...	...	...	...
V. STOCK VARIATION (total)	- 5	- 28	- 17	38	65	11	10
1. Iron and steel works	- 5	- 6	- 17	- 1	39	5	2
2. Steel foundries (independent)	...	...	...	...	...	...	...
VI. FOREIGN TRADE (balance)	- 413	- 817	- 629	- 633	- 917	-1 378	-1 521
1. Imports (-)	- 424	- 818	- 629	- 677	- 967	-1 402	-1 554
2. Exports (+)	11	1	-	44	50	24	33
A. TOTAL NATIONAL RESOURCES (utilized)(II + III + VI)	1 292	832	764	735	747	657	567
B. DOMESTIC SCRAP (total) (A - II.1)(collection utilized)	449	296	277	270	255	192	177
C. DOMESTIC SCRAP MARKETED (B - II.2) or (A - II)	181	86	97	156	146	113	97
D. EXTERNAL PURCHASE REQUIREMENTS (GROSS)(I - II) (all consumers)	466	879	720	699	909	1 414	1 541
E. EXTERNAL PURCHASE REQUIREMENTS (NET) (I - II + IV) (all consumers)	599	931	743	751	998	1 480	1 608
F. SURPLUS (+) or DEFICIT (-) (balance) (C - E) or (VI + V)	- 418	- 845	- 646	- 595	- 852	-1 367	-1 511

Source: Government replies to UN/ECE questionnaires.

Notes: Stock variation calculated as follows: V = II + III - I - IV.

1980 and 1985 pig iron data also include blast furnace ferro-alloys.

MEXICO

(Thousands of tonnes)

	1980	1985	1990	1993	1994	1995	1996	1997 Forecast
Basic Iron and Steel Data								
1. Pig iron production	3 639*	3 595*	3 665	6 159	6 717	4 142	4 240	4 339
2. DRI production	...	1 500*	2 525	2 736*	3216*	3 691	3 794	3 901
3. Crude steel production	7 156	7 399	8 734	9 198	10 246	12 128	13 169	14 000
4. CC ratio (%)	29	51	63	76*	80*	73	78	81
Iron and Steel Scrap Data								
I. CONSUMPTION (total)	...	3 487	2 277	2 732	...	3 495	4 010	4 451
1. Iron and steel works	...	1 963	2 277	2 550	...	3 495	4 010	4 451
2. Steel foundries	...	1 445	...	...	...	...	...	...
3. Iron foundries	...	...	...	...	...	...	...	...
4. Re-rolling	...	...	...	...	...	...	...	...
5. Independent electric ferro-alloy	...	...	...	182	...	...	...	...
II. OWN RESOURCES (total)	...	1 202	...	...	...	...	...	...
1. Circulating scrap	...	1 202	...	...	...	...	...	...
(a) Iron and steel works	...	1 110	...	...	...	...	...	...
(b) Steel foundries (independent)	...	92	...	...	...	...	...	...
(c) Iron foundries	...	...	...	...	...	...	...	...
2. Capital scrap (demolition, etc.)	...	-	...	...	...	...	...	...
III. EXTERNAL RECEIPTS (tota	...	2 201	...	...	...	...	...	...
1. Iron and steel works	...	769	...	...	...	...	...	...
(a) Trade	...	-	...	...	...	...	...	...
(b) Direct purchase	...	769	...	...	...	...	...	...
(c) Imports (not included in (a) or (b	...	-	...	...	...	...	...	...
2. Steel foundries (independent)	...	1 353	...	...	...	...	...	...
3. Iron foundries	...	...	...	...	...	...	...	...
4. Independent electric ferro-alloy	...	...	...	...	...	...	...	...
IV. OUTSIDE SALES (-) (total)	...	84	...	...	...	...	...	...
1. Iron and steel works	...	84	...	...	...	...	...	...
2. Steel foundries (independent)	...	-	...	...	...	...	...	...
3. Iron foundries	...	...	...	...	...	...	...	...
V. STOCK VARIATION (total)	...	- 168	...	...	...	...	...	...
1. Iron and steel works	...	- 168	...	...	...	...	...	...
2. Steel foundries (independent)	...	-	...	...	...	...	...	...
VI. FOREIGN TRADE (balance)	...	- 659	- 617	...	...	- 401	- 512	- 634
1. Imports (-)	...	- 659	- 660	...	...	- 602	- 681	- 771
2. Exports (+)	...	-	43	...	...	201	169	137
A. TOTAL NATIONAL RESOURCES (utilized)(II + III + VI)	...	2 744	...	...	...	...	...	...
B. DOMESTIC SCRAP (total) (A - II.1)(collection utilized)	...	1 542	...	...	...	...	...	...
C. DOMESTIC SCRAP MARKETE (B - II.2) or (A - II)	...	1 542	...	...	...	...	...	...
D. EXTERNAL PURCHASE REQU (GROSS)(I - II) (all consumers)	...	2 285	...	...	...	...	...	...
E. EXTERNAL PURCHASE REQU (NET) (I - II + IV) (all consumers	...	2 369	...	...	...	...	...	...
F. SURPLUS (+) or DEFICIT (-) (ba (C - E) or (VI + V)	...	- 827	...	...	...	...	...	...

Sources: Government replies to UN/ECE questionnaires; * International Iron and Steel Institute.

Notes: Stock variation calculated as follows: V = II + III - I - IV.

1996 figures are preliminary.

NETHERLANDS

(Thousands of tonnes)

	1980	1985	1990	1993	1994	1995	1996
Basic Iron and Steel Data							
1. Pig iron production	4 328	4 852	4 960	5 404	5 443	5 530	5 544
2. DRI production	...	...	...	...	...	...	...
3. Crude steel production	5 264	5 517	5 412	6 000	6 171	6 409	6 326
4. CC ratio (%)	6	39	94	97	96	96	98
Iron and Steel Scrap Data							
I. CONSUMPTION (total)	1 621	1 504	955	1 127	1 187	1 292	...
1. Iron and Steel Works	1 486	1 350	955	1 127	1 187	1 292	...
2. Steel foundries	-	...	...	...	...	...	...
3. Iron foundries	135	154	...	...	...	...	...
4. Re-rolling	...	...	...	...	...	...	...
5. Independent electric ferro-alloy	...	...	...	...	...	...	...
II. OWN RESOURCES (total)	1 043	1 102	534	550	602	543	...
1. Circulating scrap	1 043	1 102	534	550	602	543	...
(a) Iron and steel works	962	1 031	534	550	602	543	...
(b) Steel foundries (independent)	4	-	...	...	...	...	...
(c) Iron foundries	77	71	...	...	...	...	...
2. Capital scrap (demolition, etc.)	...	...	...	...	...	...	...
III. EXTERNAL RECEIPTS (tota	507	400	493	648	667	782	900
1. Iron and steel works	503	400	493	648	667	782	900
(a) Trade	503	400	...	...	...	...	...
(b) Direct purchase	...	...	...	...	...	...	...
(c) Imports (not included in (a) or (b	...	...	...	...	...	...	...
2. Steel foundries (independent)	4	-	...	...	...	...	...
3. Iron foundries	...	...	...	...	...	...	...
4. Independent electric ferro-alloy	...	...	...	...	...	...	...
IV. OUTSIDE SALES (-) (total)	...	...	...	...	...	...	...
1. Iron and steel works	...	...	...	...	...	...	...
2. Steel foundries (independent)	...	...	...	...	...	...	...
3. Iron foundries	...	...	...	...	...	...	...
V. STOCK VARIATION (total)	...	...	...	...	...	...	...
1. Iron and steel works	- 21	81	71	71	82	33	145
2. Steel foundries (independent)	6	-	...	...	...	...	...
VI. FOREIGN TRADE (balance)	1 041	1 249	1 434	1 787	760	...	...
1. Imports (-)	- 153	- 586	-1 160	-3 745	-2 935	...	...
2. Exports (+)	1 194	1 835	2 594	5 532	3 695	...	...
A. TOTAL NATIONAL RESOURCES							
(utilized)(II + III + VI)	2 591	2 751	2 461	2 985	2 029	...	...
B. DOMESTIC SCRAP (total)							
(A - II.1)(collection utilized)	...	...	...	...	...	...	...
C. DOMESTIC SCRAP MARKETE							
(B - II.2) or (A - II)	1 548	1 649	1 927	2 435	1 427	...	...
D. EXTERNAL PURCHASE REQUIREMENTS							
(GROSS)(I - II) (all consumers)	578	402	421	577	585	749	...
E. EXTERNAL PURCHASE REQUIREMENTS							
(NET) (I - II + IV) (all consumers)	...	...	...	...	...	...	...
F. SURPLUS (+) or DEFICIT (-) (balance)							
(C - E) or (VI + V)	...	...	...	...	...	...	...

Source : Government replies to UN/ECE questionnaires.

Note: 1980 and 1985 pig iron data also include blast furnace ferro-alloys.

NORWAY

(Thousands of tonnes)

	1980	1985	1990	1993	1994	1995	1996
Basic Iron and Steel Data							
1. Pig iron production	618	610	54*	73*	70*	70*	70*
2. DRI production (sponge iron)	...	...	-	-	-	-	
3. Crude steel production	866	930	376	505	456	503	530
4. CC ratio (%)	13*	56*	98	99	99	99	99
Iron and Steel Scrap Data							
I. CONSUMPTION (total)	...	526	481	614	557	582	...
1. Iron and steel works	...	475	423	565	509	536	...
2. Steel foundries	...	...	48	39	35	35	...
3. Iron foundries	...	51	...	...	...	...	...
4. Re-rolling	...	...	...	...	...	...	...
5. Independent electric ferro-alloy	...	...	10	10	13	11	...
II. OWN RESOURCES (total)	...	130	108	85	72	56	...
1. Circulating scrap	...	130	47	45	42	41	...
(a) Iron and steel works	...	120	47	45	42	41	...
(b) Steel foundries (independent)	...	-	...	...	...	...	...
(c) Iron foundries	...	10	...	...	...	...	...
2. Capital scrap (demolition, etc.)	...	-	61	40	30	15	...
III. EXTERNAL RECEIPTS (total)	...	376	312	481	437	480	...
1. Iron and steel works	...	312	312	481	437	480	...
(a) Trade	...	-	298	265	255	250	...
(b) Direct purchase	...	302	-	-	-	-	...
(c) Imports (not included in (a) or (b))	...	10	14	216	182	230	...
2. Steel foundries (independent)	...	4	...	...	...	...	...
3. Iron foundries	...	60	...	...	...	...	...
4. Independent electric ferro-alloy	...	...	...	...	...	...	...
IV. OUTSIDE SALES (-) (total)	...	-					...
1. Iron and steel works	...	-	-	-	-	-	...
2. Steel foundries (independent)	...	-	...	...	...	...	...
3. Iron foundries	...	...	...	...	...	...	...
V. STOCK VARIATION (total)	...	-20	-61	-48	-48	-46	...
1. Iron and steel works	...	-43	...	...	...	...	...
2. Steel foundries (independent)	...	4	...	...	...	...	...
VI. FOREIGN TRADE (balance)	...	-6	-10	-213	-137	-134	-71
1. Imports (-)	...	-6	-23	-223	-200	-231	-216
2. Exports (+)	...	-	12	10	63	97	145
A. TOTAL NATIONAL RESOURCES (utilized)(II + III + VI)	...	500	410	353	372	403	...
B. DOMESTIC SCRAP (total) (A - II.1)(collection utilized)	...	370	363	308	330	362	...
C. DOMESTIC SCRAP MARKETED (B - II.2) or (A - II)	...	370	302	268	300	347	...
D. EXTERNAL PURCHASE REQUIREMENTS (GROSS)(I - II) (all consumers)	...	396	373	529	485	526	...
E. EXTERNAL PURCHASE REQUIREMENTS (NET) (I - II + IV) (all consumers)	...	396	373	529	485	526	...
F. SURPLUS (+) or DEFICIT (-) (balance) (C - E) or (VI + V)	...	-26	-71	-261	-185	-180	...

Sources: Government replies to UN/ECE questionnaires; * International Iron and Steel Institute.

Notes: Stock variation calculated as follows: V = II + III - I - IV.

1980 and 1985 pig iron data also include blast furnace ferro-alloys.

POLAND

(Thousands of tonnes)

	1980	1985	1990	1993	1994	1995	1996
Basic Iron and Steel Data							
1. Pig iron production	11 510	9 436	8 423	6 162	6 932	7 497	6 581
2. DRI production	...	...	...	...	...	...	...
3. Crude steel production	19 485	16 126	13 625	9 906	11 024	11 890	10 433
4. CC ratio (%)	4	10	8	10	12	21	55
Iron and Steel Scrap Data							
I. CONSUMPTION (total)	11 864	8 744	...	5 413	5 917	...	...
1. Iron and steel works	10 629	8 744	...	5 413	5 917	...	...
2. Steel foundries	408	...	...	...	...	...	...
3. Iron foundries	819	...	...	...	...	...	...
4. Re-rolling	...	...	...	...	...	...	...
5. Independent electric ferro-alloy	...	...	...	...	...	...	...
II. OWN RESOURCES (total)	7 637	5 389	...	2 853	3 016	...	...
1. Circulating scrap	5 894	4 180	...	...	...	...	...
(a) Iron and steel works	5 361	4 180	...	...	...	...	...
(b) Steel foundries (independent)	104	...	...	...	...	...	...
(c) Iron foundries	429	...	...	...	...	...	...
2. Capital scrap (demolition, etc.)	1 743	1 209	...	...	...	...	...
III. EXTERNAL RECEIPTS (total)	4 939	3 650	...	2 699	3 428	...	...
1. Iron and steel works	4 297	3 650	...	2 699	3 428	...	...
(a) Trade	4 297	2 350	...	...	...	...	...
(b) Direct purchase	...	1 300	...	...	...	...	...
(c) Imports (not included in (a) or (b))	...	...	...	...	...	...	...
2. Steel foundries (independent)	311	...	...	...	...	...	...
3. Iron foundries	323	...	...	...	...	...	...
4. Independent electric ferro-alloy	...	...	...	...	...	...	...
IV. OUTSIDE SALES (-) (total)	724	269	...	...	227	...	...
1. Iron and steel works	627	269	...	...	227	...	...
2. Steel foundries (independent)	32	...	...	...	...	...	...
3. Iron foundries	65	...	...	...	...	...	...
V. STOCK VARIATION (total)	- 12	26	...	...	300	...	...
1. Iron and steel works	- 12	26	...	...	...	...	...
2. Steel foundries (independent)	-	...	...	...	...	...	...
VI. FOREIGN TRADE (balance)	- 153	- 3	424	1 011	448	438	415
1. Imports (-)	- 163	- 3	- 6	0	0	0	0
2. Exports (+)	10	-	429	1 012	449	438	416
A. TOTAL NATIONAL RESOURCES (utilized)(II + III + VI)	12 423	9 036	...	6 563	6 892	...	...
B. DOMESTIC SCRAP (total) (A - II.1)(collection utilized)	6 529	4 856	...	...	...	...	...
C. DOMESTIC SCRAP MARKED (B - II.2) or (A - II)	4 786	3 647	...	3 710	3 876	...	...
D. EXTERNAL PURCHASE REQUIREMENTS (GROSS)(I - II) (all consumers)	4 227	3 355	...	2 560	2 901	...	...
E. EXTERNAL PURCHASE REQUIREMENTS (NET) (I - II + IV) (all consumers)	4 951	3 624	...	...	3 128	...	...
F. SURPLUS (+) or DEFICIT (-) (balance) (C - E) or (VI + V)	- 165	23	...	...	748	...	...

Source: Government replies to UN/ECE questionnaires.

Notes: Stock variation calculated as follows: V = II + III - I - IV.

1980 and 1985 pig iron data also include blast furnace ferro-alloys.

PORTUGAL

(Thousands of tonnes)

	1980	1985	1990	1993	1994	1995	1996
Basic Iron and Steel Data							
1. Pig iron production	370	423	339	398	414	415	454
2. DRI production	...	...	...	...	...	...	...
3. Crude steel production	663	665	744	775	723	796	852
4. CC ratio (%)	42*	43*	54	99	100	100	96
Iron and Steel Scrap Data							
I. CONSUMPTION (total)	...	343	610	592	418	598	533
1. Iron and steel works	...	343	510	509	418	598	533
2. Steel foundries	...	...	17	12	...	...	...
3. Iron foundries	...	...	83	71	...	...	...
4. Re-rolling	...	...	-	-	-	-	...
5. Independent electric ferro-alloy	...	...	-	-	-	-	...
II. OWN RESOURCES (total)	...	119	172	129	118	260	118
1. Circulating scrap	...	109	151	106	109	...	...
(a) Iron and steel works	...	109	151	106	109	...	...
(b) Steel foundries (independent)	...	...	-	-	...	...	...
(c) Iron foundries	...	...	-	-	...	...	...
2. Capital scrap (demolition, etc.)	...	10	21	23	9	...	...
III. EXTERNAL RECEIPTS (tota	...	205	313	453	362	349	400
1. Iron and steel works	...	205	313	396	362	349	400
(a) Trade	...	143	238	377	335	289	365
(b) Direct purchase	...	-	...	...	...	...	...
(c) Imports (not included in (a) or (b	...	62	75	19	27	60	35
2. Steel foundries (independent)	...	...	-	-	...	...	...
3. Iron foundries	...	...	...	57	...	...	...
4. Independent electric ferro-alloy	...	...	...	-	-	...	...
IV. OUTSIDE SALES (-) (total)	...	1					
1. Iron and steel works	...	1	-	-	-	-	-
2. Steel foundries (independent)	...	...	-	-	...	...	...
3. Iron foundries	...	...	-	-	...	...	...
V. STOCK VARIATION (total)	...	-20	-125	-10	62	11	-15
1. Iron and steel works	...	-20	-25	...	...	11	- 15
2. Steel foundries (independent)	...	...	-	-	...	...	...
VI. FOREIGN TRADE (balance)	...	-62	-68	40	87	31	23
1. Imports (-)	...	-62	-83	-23	-36	-60	-35
2. Exports (+)	...	-	15	63	123	91	58
A. TOTAL NATIONAL RESOURCES (utilized)(II + III + VI)	...	262	417	622	567	640	541
B. DOMESTIC SCRAP (total) (A - II.1)(collection utilized)	...	153	266	516	458	...	...
C. DOMESTIC SCRAP MARKETED (B - II.2) or (A - II)	...	143	245	493	449	380	423
D. EXTERNAL PURCHASE REQUIREMENTS (GROSS)(I - II) (all consumers)	...	224	438	463	300	338	415
E. EXTERNAL PURCHASE REQUIREMENTS (NET) (I - II + IV) (all consumers)	...	225	438	463	300	338	415
F. SURPLUS (+) or DEFICIT (-) (balance) (C - E) or (VI + V)	...	-82	- 193	30	149	42	8

Sources: Government replies to UN/ECE questionnaires; * International Iron and Steel Institute.

Notes: Stock variation calculated as follows: V = II + III - I - IV.

1980 and 1985 pig iron data also include blast furnace ferro-alloys.

REPUBLIC OF KOREA

(Thousands of tonnes)

	1980	1985	1990	1993	1994	1995	1996	1997 Forecast
Basic Iron and Steel Data								
1. Pig iron production	5 577*	8 833*	15 195	21 455	20 892	21 975	22 628	22 342
2. DRI production	...	...	...	...	...	...	...	...
3. Crude steel production	8 558*	13 539*	23 125	33 026	33 745	36 772	38 903	42 200
4. CC ratio (%)	32*	63*	96	98	98	98	98	98
Iron and Steel Scrap Data								
I. CONSUMPTION (total)	...	...	11 613	14 708	16 222	19 164	20 854	24 350
1. Iron and steel works	...	...	9 764	13 214	14 638	17 392	19 201	22 620
2. Steel foundries	...	...	...	...	...	...	...	...
3. Iron foundries	...	...	1 425	1 219	1 298	1 418	1 422	1 530
4. Re-rolling	...	...	424	275	286	354	231	200
5. Independent electric ferro-alloy	...	...	...	...	...	...	...	...
II. OWN RESOURCES (total)	...	...	2 267	2 801	2 973	2 988	3 022	3 270
1. Circulating scrap	...	...	2 267	2 801	2 973	2 988	3 022	3 270
(a) Iron and steel works	...	...	1 960	2 497	2 664	2 660	2 679	2 920
(b) Steel foundries (independent)	...	...	...	...	...	...	...	...
(c) Iron foundries	...	...	307	304	309	328	343	350
2. Capital scrap (demolition, etc.)	...	...	-	-	-	-	-	-
III. EXTERNAL RECEIPTS (total)	...	...	8 340	11 689	12 974	14 445	15 668	18 280
1. Iron and steel works	...	...	7 259	10 774	11 985	13 474	14 607	17 130
(a) Trade	...	...	-	...	...	-	-	-
(b) Direct purchase	...	...	3 581	5 939	7 179	8 642	9 561	10 720
(c) Imports (not included in (a) or (b))	...	...	3 678	4 835	4 806	4 832	5 046	6 410
2. Steel foundries (independent)	...	...	...	...	...	...	...	...
3. Iron foundries	...	...	1 081	915	989	971	1 061	1 150
4. Independent electric ferro-alloy	...	...	...	...	...	...	...	...
IV. OUTSIDE SALES (-) (total)	...	...	38	...	...	50	12	10
1. Iron and steel works	...	...	38	...	...	21	12	10
2. Steel foundries (independent)	...	...	...	...	...	...	...	...
3. Iron foundries	...	...	-	...	...	29	-	-
V. STOCK VARIATION (total)	...	...	-1 044	...	...	-1 781	-2 176	-2 810
1. Iron and steel works	...	...	601	...	...	905	605	705
2. Steel foundries (independent)	...	...	...	...	...	...	...	...
VI. FOREIGN TRADE (balance)	...	...	-3 845	-4 893	-4 957	-5 006	-5 100	-6 490
1. Imports (-)	...	...	-3 885	-4 903	-4 967	-5 027	-5 115	-6 500
2. Exports (+)	...	...	40	10	10	21	15	10
A. TOTAL NATIONAL RESOURCES (utilized)(II + III + VI)	...	...	6 762	9 597	10 990	12 427	13 590	15 060
B. DOMESTIC SCRAP (total) (A - II.1)(collection utilized)	...	...	4 495	6 796	8 017	9 439	10 568	11 790
C. DOMESTIC SCRAP MARKETED (B - II.2) or (A - II)	...	...	4 495	6 796	8 017	9 439	10 568	11 790
D. EXTERNAL PURCHASE REQUIREMENTS (GROSS)(I - II) (all consumers)	...	...	9 346	11 907	13 249	16 176	17 832	21 080
E. EXTERNAL PURCHASE REQUIREMENTS (NET) (I - II + IV) (all consumers)	...	...	9 384	...	...	16 226	17 844	21 090
F. SURPLUS (+) or DEFICIT (-) (balance) (C - E) or (VI + V)	...	...	-4 889	...	...	-6 787	-7 276	-9 300

Sources: Government replies to UN/ECE questionnaires; * International Iron and Steel Institute.

Note: Stock variation calculated as follows: V = II + III - I - IV.

ROMANIA

(Thousands of tonnes)

	1980	1985	1990	1992	1993	1994	1995	1996
Basic Iron and Steel Data								
1. Pig iron production	9 102*	9 212*	6 355	3 111	3 189	3 504	4 203	4 025
2. DRI production	...	...	...	...	...	...	-	-
3. Crude steel production	13 175*	13 795*	9 754	5 376	5 446	5 800	6 557	6 082
4. CC ratio (%)	18*	30*	36	42	50	47	48	49
Iron and Steel Scrap Data								
I. CONSUMPTION (total)	...	...	...	3 459	3 265	3 397	3 519	...
1. Iron and steel works	...	...	...	3 459	3 265	3 397	3 519	...
2. Steel foundries	...	...	-	-	-	-	-	...
3. Iron foundries	...	...	-	-	-	-	-	...
4. Re-rolling	...	...	-	-	-	-	-	...
5. Independent electric ferro-alloy	...	...	-	-	-	-		
II. OWN RESOURCES (total)	...	...	...	2 045	2 103	2 325	2 251	...
1. Circulating scrap	...	...	...	1 441	1 498	1 786	2 022	...
(a) Iron and steel works	...	...	...	1 441	1 498	1 786	2 022	...
(b) Steel foundries (independent)	...	...	-	-	-	-		...
(c) Iron foundries	...	...	-	-	-	-	-	...
2. Capital scrap (demolition, etc.)	...	...	...	604	605	539	229	...
III. EXTERNAL RECEIPTS (to	...	...	...	1 512	1 653	1 768	1 743	...
1. Iron and steel works	...	...	...	1 512	1 653	1 768	1743	...
(a) Trade	...	...	...	964	1 051	727	556	...
(b) Direct purchase	...	...	...	548	602	1 041	1187	...
(c) Imports (not included in (a) or	...	...	-	-	-	-	-	...
2. Steel foundries (independent)	...	...	-	-	-	-		...
3. Iron foundries	...	...	-	-	-	-	-	...
4. Independent electric ferro-alloy	...	...	-	-	-	-	-	...
IV. OUTSIDE SALES (-) (total)	...	...	...	625	579	662	485	...
1. Iron and steel works	...	...	...	625	579	662	485	...
2. Steel foundries (independent)	...	...	-	-	-	-		...
3. Iron foundries	...	...	-	-	-	-	-	...
V. STOCK VARIATION (total)	...	...	...	-527	- 88	34	- 10	...
1. Iron and steel works	...	...	...	- 527	- 88	34	-10	...
2. Steel foundries (independent)	...	...	-	-	-	-		...
VI. FOREIGN TRADE (balance	...	...	-	134	16	5	8	27
1. Imports (-)	...	...			- 1	-	-	-
2. Exports (+)	...	...	-	134	17	5	8	27
A. TOTAL NATIONAL RESOURCES (utilized)(II + III + VI)	...	...	...	3 691	3 772	4 098	4002	...
B. DOMESTIC SCRAP (total) (A - II.1)(collection utilized)	...	...	...	2 250	2 274	2 312	1980	...
C. DOMESTIC SCRAP MARKETED (B - II.2) or (A - II)	...	...	...	1 646	1 669	1 773	1751	...
D. EXTERNAL PURCHASE REQUIREMENTS (GROSS)(I - II) (all consumers	...	...	...	1 414	1 162	1 072	1268	...
E. EXTERNAL PURCHASE REQUIREMENTS (NET) (I - II + IV) (all consume	...	...	...	2 039	1 741	1 734	1753	...
F. SURPLUS (+) or DEFICIT (-) (balance) (C - E) or (VI + V)	...	...	...	- 393	- 72	39	-2	...

Sources: Government replies to UN/ECE questionnaires; * International Iron and Steel Institute.

Note: Stock variation calculated as follows: V = II + III - I - IV.

RUSSIAN FEDERATION

(Thousands of tonnes)

	1980	1985	1990	1993	1994	1995	1996	1997 Forecast
Basic Iron and Steel Data								
1. Pig iron production	107 282	109 980	59 232	40 818	36 057	38 600	36 370	38 900
2. DRI production	...	...	1 690	1 544	1 708	1 700	1 700	1 700
3. Crude steel production	147 931	154 653	81 434	54 896	46 648	51 336	49 161	51 400
4. CC ratio (%)	11	14	25	29	33	37	41	41
Iron and Steel Scrap Data								
I. CONSUMPTION (total)	...	90 100	53 842	31 356	24 469	27 925	26 903	27 823
1. Iron and steel works	...	66 000	40 313	25 193	20 307	23 987	23 080	24 000
2. Steel foundries	...	11 900	7 203	3 241	2 063	2 070	2 053	2 053
3. Iron foundries	...	11 100	6 013	2 761	2 015	1 783	1 770	1 770
4. Re-rolling	...	...	-	-	-	-	-	-
5. Independent electric ferro-alloy	...	...	313	161	84	85	-	-
II. OWN RESOURCES (total)	...	...	28 798	18 495	15 206	14 833	13 940	14 940
1. Circulating scrap	...	34 900	25 033	15 706	13 164	11 318	10 940	11 440
(a) Iron and steel works	...	31 400	20 110	13 330	11 349	9 563	9 200	9 700
(b) Steel foundries (independent)	...	3 500	1 947	899	736	890	880	880
(c) Iron foundries	...	...	2 976	1 477	1 079	865	860	860
2. Capital scrap (demolition, etc.)	...	...	3 765	2 789	2 042	3 515	3 000	3 500
III. EXTERNAL RECEIPTS (total)	...	46 300	29 078	14 713	10 060	19 241	19 183	19 183
1. Iron and steel works	...	35 600	20 472	10 926	7 713	17 007	17 100	17 100
(a) Trade	...	35 600	19 741	7 197	...	9 100	9 100	9 100
(b) Direct purchase	...	-	731	3 597	...	7 750	7 840	7 840
(c) Imports (not included in (a) or (b))	...	-	...	132	100	157	160	160
2. Steel foundries (independent)	...	9 600	5 256	2 342	1 327	1 231	1 173	1 173
3. Iron foundries	...	-	3 037	1 284	936	918	910	910
4. Independent electric ferro-alloy	...	...	313	161	84	85	-	-
IV. OUTSIDE SALES (-) (total)	...	...	3 986	1 814	1 120	6 282	6 300	6 300
1. Iron and steel works	...	8 200	3 986	1 814	1 120	6 282	6 300	6 300
2. Steel foundries (independent)	...	...	...	...	...	...	...	...
3. Iron foundries	...	...	...	...	...	...	...	...
V. STOCK VARIATION (total)	...	...	48	38	- 323	- 133	- 80	...
1. Iron and steel works	...	300	- 252	38	- 323	- 178	- 80	-
2. Steel foundries (independent)	...	100	...	...	...	...	...	...
VI. FOREIGN TRADE (balance)	...	3 315	...	302	380	12	10	10
1. Imports (-)	...	-	...	- 140	- 100	- 157	- 160	- 160
2. Exports (+)	...	3 315	832	442	480	169	170	170
A. TOTAL NATIONAL RESOURCES (utilized)(II + III + VI)	...	...	...	33 510	25 646	34 086	33 133	34 133
B. DOMESTIC SCRAP (total) (A - II.1)(collection utilized)	...	...	...	17 804	12 482	22 768	22 193	22 693
C. DOMESTIC SCRAP MARKETED (B - II.2) or (A - II)	...	...	...	15 015	10 440	19 253	19 193	19 193
D. EXTERNAL PURCHASE REQUIREMENTS (GROSS)(I - II) (all consumers)	...	...	25 044	12 861	9 263	13 092	12 963	12 883
E. EXTERNAL PURCHASE REQUIREMENTS (NET) (I - II + IV) (all consumers)	...	...	29 030	14 675	10 383	19 374	19 263	19 183
F. SURPLUS (+) or DEFICIT (-) (balance) (C - E) or (VI + V)	...	...	...	340	57	- 121	- 70	...

Source: Government replies to UN/ECE questionnaires.

Notes: Stock variation calculated as follows: V = II + III - I - IV.

All data available for 1980 and 1985 represent the data of USSR as communicated.

SLOVAKIA

(Thousands of tonnes)

	1985	1990	1992	1993	1994	1995	1996	1997 Forecast
Basic Iron and Steel Data								
1. Pig iron production	...	3 561	2 952	3 205	3 330	3 207	2 928	2 990
2. DRI production	...	...	...	...	...	-	-	-
3. Crude steel production	...	4 681	3 597	3 769	3 805	3 933	3 582	3 700
4. CC ratio (%)	...	26	44	72	67	89	100	100
Iron and Steel Scrap Data								
I. CONSUMPTION (total)	...	1 777	1 371	1 284	1 259	1 195	1 343	1 340
1. Iron and steel works	...	1 347	1 084	1 054	997	1 108	1 260	1 250
2. Steel foundries	...	74	50	40	32	42	41	45
3. Iron foundries	...	73	76	64	51	29	22	25
4. Re-rolling	...	255	149	113	165	-	-	-
5. Independent electric ferro-alloy	...	28	12	13	14	16	20	20
II. OWN RESOURCES (total)	...	799	649	489	574	463	392	424
1. Circulating scrap	...	751	555	467	486	463	392	424
(a) Iron and steel works	...	716	527	445	466	440	369	400
(b) Steel foundries (independent)	...	14	8	7	7	10	11	11
(c) Iron foundries	...	21	20	15	14	13	12	13
2. Capital scrap (demolition, etc.)	...	48	94	22	88	-	-	-
III. EXTERNAL RECEIPTS (tota	...	602	580	622	536	716	931	896
1. Iron and steel works	...	490	482	540	474	668	891	850
(a) Trade	...	490	482	540	455	433	733	650
(b) Direct purchase	...	-	-	-	19	-	-	-
(c) Imports (not included in (a) or (b	...	-	-	-	-	235	158	200
2. Steel foundries (independent)	...	60	41	33	25	32	30	34
3. Iron foundries	...	52	56	49	37	16	10	12
4. Independent electric ferro-alloy	...	28	12	13	14	16	20	20
IV. OUTSIDE SALES (-) (total)	...							
1. Iron and steel works	...	-	-	-	-	-	-	-
2. Steel foundries (independent)	...	-	-	-	-	-	-	-
3. Iron foundries	...	-	-	-	-	-	-	-
V. STOCK VARIATION (total)	...	- 376	- 142	- 173	- 148	-16	-20	-20
1. Iron and steel works	...	...	...	...	...	...	...	...
2. Steel foundries (independent)	...	...	...	...	...	...	...	...
VI. FOREIGN TRADE (balance)	...	-	205	174	162	-38	54	-
1. Imports (-)	...	-	...	- 30	- 40	-235	-158	-200
2. Exports (+)	...	-	205	204	202	197	212	200
A. TOTAL NATIONAL RESOURCES (utilized)(II + III + VI)	...	1 401	1 434	1 285	1 273	1141	1 377	1 320
B. DOMESTIC SCRAP (total) (A - II.1)(collection utilized)	...	650	879	818	786	678	985	896
C. DOMESTIC SCRAP MARKETED (B - II.2) or (A - II)	...	602	785	796	698	678	985	896
D. EXTERNAL PURCHASE REQUIREMENTS (GROSS)(I - II) (all consumers)	...	978	722	795	684	732	951	916
E. EXTERNAL PURCHASE REQUIREMENTS (NET) (I - II + IV) (all consumers)	...	978	722	795	684	732	951	916
F. SURPLUS (+) or DEFICIT (-) (balance) (C - E) or (VI + V)	...	- 376	63	1	14	-54	34	-20

Source: Government replies to UN/ECE questionnaires.

Note: Stock variation calculated as follows: V = II + III - I - IV.

SOUTH AFRICA

(Thousands of tonnes)

	1980	1985	1990	1993	1994	1995	1996
Basic Iron and Steel Data							
1. Pig iron production	7 218*	6 010*	6 257	6 121	6 047	6224*	6014*
2. DRI production	...	543*	882	819	935	913*	960*
3. Crude steel production	9 067*	8 507*	8 620	8 726	8 525	8741*	7973*
4. CC ratio (%)	52*	65*	73	85	91	92*	94*
Iron and Steel Scrap Data							
I. CONSUMPTION (total)	...	...	3 229	3 514	2 971	2 754	...
1. Iron and steel works	...	...	3 229	3 514	2 971	2 754	...
2. Steel foundries	...	...	...	...	...	...	...
3. Iron foundries	...	...	...	...	...	...	...
4. Re-rolling	...	...	...	...	...	...	...
5. Independent electric ferro-alloy	...	...	...	...	...	...	...
II. OWN RESOURCES (total)	...	...	3 043	3 114	2 657	2 529	...
1. Circulating scrap	...	...	...	...	...	...	...
(a) Iron and steel works	...	...	...	...	...	...	...
(b) Steel foundries (independent)	...	...	...	...	...	...	...
(c) Iron foundries	...	...	...	...	...	...	...
2. Capital scrap (demolition, etc.)	...	...	...	...	...	...	...
III. EXTERNAL RECEIPTS (total)	...	...	...	...	...	...	...
1. Iron and steel works	...	...	...	...	...	...	...
(a) Trade	...	...	...	...	...	...	...
(b) Direct purchase	...	...	...	...	...	...	...
(c) Imports (not included in (a) or (b))	...	...	...	...	...	...	...
2. Steel foundries (independent)	...	...	...	...	...	...	...
3. Iron foundries	...	...	...	...	...	...	...
4. Independent electric ferro-alloy	...	...	...	...	...	...	...
IV. OUTSIDE SALES (-) (total)	...	...	...	...	...	...	...
1. Iron and steel works	...	...	...	...	...	...	...
2. Steel foundries (independent)	...	...	...	...	...	...	...
3. Iron foundries	...	...	...	...	...	...	...
V. STOCK VARIATION (total)	...	...	...	...	...	...	...
1. Iron and steel works	...	...	-	-	-	-	...
2. Steel foundries (independent)	...	...	...	...	...	...	...
VI. FOREIGN TRADE (balance)	...	...	71	80	78	129	...
1. Imports (-)	...	...	...	-32	-99	-15	...
2. Exports (+)	...	...	71	111	177	144	...
A. TOTAL NATIONAL RESOURCES (utilized)(II + III + VI)	...	...	...	...	...	...	...
B. DOMESTIC SCRAP (total) (A - II.1)(collection utilized)	...	...	...	...	...	...	...
C. DOMESTIC SCRAP MARKETED (B - II.2) or (A - II)	...	...	...	...	...	...	...
D. EXTERNAL PURCHASE REQUIREMENTS (GROSS)(I - II) (all consumers)	...	...	186	400	314	225	...
E. EXTERNAL PURCHASE REQUIREMENTS (NET) (I - II + IV) (all consumers)	...	...	...	...	...	...	...
F. SURPLUS (+) or DEFICIT (-) (balance) (C - E) or (VI + V)	...	...	...	...	...	...	...

Sources: Government replies to UN/ECE questionnaires; * International Iron and Steel Institute.

SPAIN

(Thousands of tonnes)

	1980	1985	1990	1993	1994	1995	1996
Basic Iron and Steel Data							
1. Pig iron production	6 379	5 457	5 646	5 394	5 447	5 106	4 127
2. DRI production (sponge iron)	...	...	-	-	-	-	-
3. Crude steel production	12 642	14 032	12 711	12 961	13 435	13 802	12 161
4. CC ratio (%)	36*	57*	89	97	96	96	96
Iron and Steel Scrap Data							
I. CONSUMPTION (total)	9 062	10 940	8 921	9 416	9 771	10 186	...
1. Iron and steel works	8 307	10 693	8 921	9 416	9 771	10 186	...
2. Steel foundries	221	...	...	...	...	...	...
3. Iron foundries	150	247	...	...	...	...	...
4. Re-rolling	...	...	...	...	...	...	...
5. Independent electric ferro-alloy	...	...	...	...	...	...	...
II. OWN RESOURCES (total)	2 198	...	1 492	1 514	1 625	1 434	...
1. Circulating scrap	2 198	...	1 492	1 514	1 625	1 434	...
(a) Iron and steel works	2 006	...	1 492	1 514	1 625	1 434	...
(b) Steel foundries (independent)	88	...	...	...	...	...	...
(c) Iron foundries	50	...	...	...	...	...	...
2. Capital scrap (demolition, etc.)	...	...	...	...	...	...	...
III. EXTERNAL RECEIPTS (total)	7 018	4 146	7 413	7 893	8 626	8 949	...
1. Iron and steel works	6 455	4 146	7 413	7 893	8 413	...	
(a) Trade	6 455	...	3 827	3 086	3 883	...	...
(b) Direct purchase	...	...	301	767	305	...	...
(c) Imports (not included in (a) or (b))	...	4 146	3 285	4 040	4 225	...	...
2. Steel foundries (independent)	133	...	...	...	...	...	...
3. Iron foundries	100	...	...	...	...	...	...
4. Independent electric ferro-alloy	...	...	...	...	...	...	...
IV. OUTSIDE SALES (-) (total)	...	...	136	128	153	...	...
1. Iron and steel works	...	...	136	128	153	...	...
2. Steel foundries (independent)	...	...	...	...	...	...	...
3. Iron foundries	...	...	...	...	...	...	...
V. STOCK VARIATION (total)	...	...	- 152	- 137	327	...	...
1. Iron and steel works	154	...	- 151	- 20	114	...	...
2. Steel foundries (independent)	-	...	...	...	...	...	...
VI. FOREIGN TRADE (balance)	-4 025	-4 146	-4 268	-3 928	-4 608	-4 864	-4 461
1. Imports (-)	-4 026	-4 146	-4 291	-3 995	-4 624	-4 883	-4 479
2. Exports (+)	1	-	23	67	16	19	18
A. TOTAL NATIONAL RESOURCES (utilized)(II + III + VI)	5 191	...	4 637	5 479	4 018	4 085	...
B. DOMESTIC SCRAP (total) (A - II.1)(collection utilized)	...	...	...	...	...	...	...
C. DOMESTIC SCRAP MARKETED (B - II.2) or (A - II)	2 993	-	3 145	3 965	2 393	2 651	...
D. EXTERNAL PURCHASE REQUIREMENTS (GROSS)(I - II) (all consumers)	6 864	...	7 429	7 902	8 146	8 752	...
E. EXTERNAL PURCHASE REQUIREMENTS (NET) (I - II + IV) (all consumers)	6 864	...	7 565	8 030	8 299	...	...
F. SURPLUS (+) or DEFICIT (-) (balance) (C - E) or (VI + V)	...	...	-4 420	-4 065	-5 906	...	...

Sources: Government replies to UN/ECE questionnaires; * International Iron and Steel Institute.

Notes: Stock variation calculated as follows: V = II + III - I - IV.

1980 and 1985 pig iron data also include blast furnace ferro-alloys.

SWEDEN

(Thousands of tonnes)

	1980	1985	1990	1993	1994	1995	1996	1997 Forecast
Basic Iron and Steel Data								
1. Pig iron production	2 435	2 424	2 736	2 845	3 037	3 020	3 129	3 130
2. DRI production (sponge iron)	...	108*	109	95	109	124	125	...
3. Crude steel production	4 237	4 705	4 454	4 591	4 967	4 953	4 910	4 950
4. CC ratio (%)	49	81	86	89	87	86	88	...
Iron and Steel Scrap Data								
I. CONSUMPTION (total)	2 788	3 313	2 168	2 282	2 683	2 466	2 395	...
1. Iron and steel works	2 504	3 066	2 168	2 282	2 683	2 466	2 395	...
2. Steel foundries	...	...	...	...	...	...	...	...
3. Iron foundries	284	247	...	...	...	...	...	...
4. Re-rolling	...	...	...	...	...	...	...	...
5. Independent electric ferro-alloy	...	...	...	...	...	...	...	...
II. OWN RESOURCES (total)	1 587	...	979	738	905	1 141	1 051	...
1. Circulating scrap	1 405	82	975	738	905	1 068	992	...
(a) Iron and steel works	1 311	...	975	738	905	1 068	992	...
(b) Steel foundries (independent)	...	...	...	...	...	...	...	...
(c) Iron foundries	94	82	...	...	...	...	...	...
2. Capital scrap (demolition, etc.)	182	...	4	...	...	73	59	...
III. EXTERNAL RECEIPTS (tota	1 108	165	247	469	662	1 262	1 224	...
1. Iron and steel works	918	...	247	469	662	1 262	1 224	...
(a) Trade	843	...	-	-	-	840	810	...
(b) Direct purchase	...	...	-	-	-	-	-	...
(c) Imports (not included in (a) or (b	75	...	247	469	662	422	414	...
2. Steel foundries (independent)	...	...	...	...	...	...	...	...
3. Iron foundries	190	165	...	...	...	...	...	...
4. Independent electric ferro-alloy	...	...	...	-	-	...	...	...
IV. OUTSIDE SALES (-) (total)	13	...	...	...	...	24	44	...
1. Iron and steel works	13	...	...	...	...	24	44	...
2. Steel foundries (independent)	...	...	...	...	...	...	...	...
3. Iron foundries	...	...	...	...	...	...	...	...
V. STOCK VARIATION (total)	- 106	...	...	...	...	- 87	- 164	...
1. Iron and steel works	- 106	...	...	...	...	...	...	...
2. Steel foundries (independent)	217	...	...	...	...	...	...	...
VI. FOREIGN TRADE (balance)	- 59	- 863	- 168	- 237	- 323	- 19	- 4	
1. Imports (-)	- 75	- 885	- 222	- 468	- 653	- 282	-269	...
2. Exports (+)	16	22	54	231	330	263	265	...
A. TOTAL NATIONAL RESOURCES (utilized)(II + III + VI)	2 636	...	1 058	970	1 480	2 384	2 271	...
B. DOMESTIC SCRAP (total) (A - II.1)(collection utilized)	1 231	...	83	...	...	1 316	1 279	...
C. DOMESTIC SCRAP MARKETED (B - II.2) or (A - II)	1 049	...	79	232	339	1 243	1 220	...
D. EXTERNAL PURCHASE REQUIREMENTS (GROSS)(I - II) (all consumers)	1 201	...	1 189	1 544	1 542	1 325	1 344	...
E. EXTERNAL PURCHASE REQUIREMENTS (NET) (I - II + IV) (all consumers)	1 214	...	...	...	...	1 349	1 388	...
F. SURPLUS (+) or DEFICIT (-) (balance) (C - E) or (VI + V)	- 165	...	...	...	...	- 106	- 168	...

Source: Government replies to UN/ECE questionnaires.

Notes: Stock variation calculated as follows: V = II + III - I - IV.

1980 and 1985 pig iron data also include blast furnace ferro-alloys.

SWITZERLAND

(Thousands of tonnes)

	1980	1985	1990	1993	1994	1995	1996
Basic Iron and Steel Data							
1. Pig iron production	33	66	129*	82*	80*	80*	80*
2. DRI production	...	...	...	...	...	...	...
3. Crude steel production	929	978	1 105	1 254	1 100	850	750
4. CC ratio (%)	33*	66*	100	100	100	100	100
Iron and Steel Scrap Data							
I. CONSUMPTION (total)	...	...	1 215	1 385	1 210	...	...
1. Iron and steel works	...	...	1 215	1 385	1 210	...	...
2. Steel foundries	...	...	...	...	...	...	...
3. Iron foundries	...	...	...	...	...	...	...
4. Re-rolling	...	...	...	...	...	...	...
5. Independent electric ferro-alloy	...	...	...	...	...	...	...
II. OWN RESOURCES (total)	...	...	65	64	55	...	...
1. Circulating scrap	...	...	65	64	55	...	...
(a) Iron and steel works	...	...	65	64	55	...	...
(b) Steel foundries (independent)	...	...	...	...	...	...	...
(c) Iron foundries	...	...	...	...	...	...	...
2. Capital scrap (demolition, etc.)	...	...	-	-	-	...	...
III. EXTERNAL RECEIPTS (total)	...	...	1 153	1 401	1 186	...	...
1. Iron and steel works	...	...	1 153	1 401	1 186	...	...
(a) Trade	...	...	-	-	-	...	...
(b) Direct purchase	...	...	1 017	972	1 014	...	...
(c) Imports (not included in (a) or (b))	...	...	136	429	172	...	...
2. Steel foundries (independent)	...	...	...	...	...	...	...
3. Iron foundries	...	...	...	...	...	...	...
4. Independent electric ferro-alloy	...	...	...	...	...	...	...
IV. OUTSIDE SALES (-) (total)	...	...	...	...	...	...	...
1. Iron and steel works	...	...	...	...	...	...	...
2. Steel foundries (independent)	...	...	...	...	...	...	...
3. Iron foundries	...	...	...	...	...	...	...
V. STOCK VARIATION (total)	...	...	...	...	...	...	...
1. Iron and steel works	...	...	...	...	...	...	...
2. Steel foundries (independent)	...	...	...	...	...	...	...
VI. FOREIGN TRADE (balance)	...	...	- 73	- 362	- 87	...	...
1. Imports (-)	...	...	- 136	- 429	- 172	...	...
2. Exports (+)	...	...	63	67	85	...	...
A. TOTAL NATIONAL RESOURCES (utilized)(II + III + VI)	...	...	1 145	1 103	1 154	...	...
B. DOMESTIC SCRAP (total) (A - II.1)(collection utilized)	...	...	1 080	1 039	1 099	...	...
C. DOMESTIC SCRAP MARKETED (B - II.2) or (A - II)	...	...	1 080	1 039	1 099	...	...
D. EXTERNAL PURCHASE REQUIREMENTS (GROSS)(I - II) (all consumers)	...	...	1 150	1 321	1 155	...	...
E. EXTERNAL PURCHASE REQUIREMENTS (NET) (I - II + IV) (all consumers)	...	...	...	...	...	...	...
F. SURPLUS (+) or DEFICIT (-) (balance) (C - E) or (VI + V)	...	...	...	...	...	...	...

Sources: Government replies to UN/ECE questionnaires; * International Iron and Steel Institute.

Notes: Stock variation calculated as follows: V = II + III - I - IV.

1980 and 1985 pig iron data also include blast furnace ferro-alloys.

THE FORMER YUGOSLAV REPUBLIC OF MACEDONIA

(Thousands of tonnes)

	1990	1991	1992	1993	1994	1995	1996
Basic Iron and Steel Data							
1. Pig iron production	...	...	108	78	63	72	...
2. DRI production	...	...	...	...	...	...	...
3. Crude steel production	208	221	171	137	67	32	2
4. CC ratio (%)	...	...	95	97	96	94	-
Iron and Steel Scrap Data							
I. CONSUMPTION (total)	...	...	103	58	37	11	...
1. Iron and steel works	...	...	103	58	37	11	...
2. Steel foundries	...	...	...	...	...	...	...
3. Iron foundries	...	...	...	...	...	...	...
4. Re-rolling	...	...	...	...	...	...	...
5. Independent electric ferro-alloy	...	...	...	...	...	...	...
II. OWN RESOURCES (total)	...	...	222	139	79	34	...
1. Circulating scrap	...	...	...	...	...	...	...
(a) Iron and steel works	...	...	...	...	...	...	...
(b) Steel foundries (independent)	...	...	...	...	...	...	...
(c) Iron foundries	...	...	...	...	...	...	...
2. Capital scrap (demolition, etc.)	...	...	...	...	...	...	...
III. EXTERNAL RECEIPTS (total)	...	...	...	...	-	-	...
1. Iron and steel works	...	...	...	...	...	...	...
(a) Trade	...	...	...	...	...	...	...
(b) Direct purchase	...	...	...	...	...	...	...
(c) Imports (not included in (a) or (b))	...	...	...	...	...	...	...
2. Steel foundries (independent)	...	...	...	...	...	...	...
3. Iron foundries	...	...	...	...	...	...	...
4. Independent electric ferro-alloy	...	...	...	...	...	...	...
IV. OUTSIDE SALES (-) (total)	...	...	...	...	...	...	...
1. Iron and steel works	...	...	...	...	...	...	...
2. Steel foundries (independent)	...	...	...	...	...	...	...
3. Iron foundries	...	...	...	...	...	...	...
V. STOCK VARIATION (total)	...	...	...	...	...	...	...
1. Iron and steel works	...	...	...	...	...	...	...
2. Steel foundries (independent)	...	...	...	...	...	...	...
VI. FOREIGN TRADE (balance)	...	...	...	-15	-	1	...
1. Imports (-)	...	...	...	-15	-	1	...
2. Exports (+)	...	...	...	-	-	-	...
A. TOTAL NATIONAL RESOURCES (utilized)(II + III + VI)	...	...	...	...	...	...	...
B. DOMESTIC SCRAP (total) (A - II.1)(collection utilized)	...	...	...	...	...	...	...
C. DOMESTIC SCRAP MARKETED (B - II.2) or (A - II)	...	...	...	...	...	...	...
D. EXTERNAL PURCHASE REQUIREMENTS (GROSS)(I - II) (all consumers)	...	...	-119	-81	-42	-23	...
E. EXTERNAL PURCHASE REQUIREMENTS (NET) (I - II + IV) (all consumers)	...	...	...	...	...	...	...
F. SURPLUS (+) or DEFICIT (-) (balance) (C - E) or (VI + V)	...	...	...	...	...	...	...

Source: Government replies to UN/ECE questionnaires.

TURKEY

(Thousands of tonnes)

	1980	1985	1990	1993	1994	1995	1996	1997 Forecast
Basic Iron and Steel Data								
1. Pig iron production	2 040	3 371	5 367	4 931	4 604	4 403	5 253	5 300
2. DRI production	...	...	-	-	-	-	-	-
3. Crude steel production	2 535	4 113	9 443	11 519	12 624	13 183	13 382	13 700
4. CC ratio (%)	42	66	92	94	91	92	93	...
Iron and Steel Scrap Data								
I. CONSUMPTION (total)	...	...	6 048	8 023	...	10 769	10 947	...
1. Iron and steel works	...	...	5 760	8 023	...	10 499	10 502	...
2. Steel foundries	...	...	...	...	...	...	...	...
3. Iron foundries	...	...	...	...	...	...	...	...
4. Re-rolling	...	...	...	...	...	...	...	...
5. Independent electric ferro-alloy	...	...	288	...	...	270	445	
II. OWN RESOURCES (total)	...	342	203	275	...	672	895	651
1. Circulating scrap	...	306	192	264	...	639	847	614
(a) Iron and steel works	...	306	192	264	...	639	847	614
(b) Steel foundries (independent)	...	...	...	...	...	...	...	...
(c) Iron foundries	...	...	...	...	...	...	...	...
2. Capital scrap (demolition, etc.)	...	36	11	11	...	33	48	37
III. EXTERNAL RECEIPTS (tota	...	...	539	...	...	1 503	1 830	1 471
1. Iron and steel works	...	...	311	...	...	1 220	1 390	1 471
(a) Trade	...	...	20	...	...	638	705	763
(b) Direct purchase	...	...	114	...	...	93	281	413
(c) Imports (not included in (a) or (b	...	...	177	...	...	489	404	295
2. Steel foundries (independent)	...	...	...	...	...	...	...	...
3. Iron foundries	...	...	...	...	...	...	...	...
4. Independent electric ferro-alloy	...	...	228	...	...	282	440	
IV. OUTSIDE SALES (-) (total)	...	...	...	...	...	...	...	...
1. Iron and steel works	...	2	6	...	...	-	-	-
2. Steel foundries (independent)	...	...	...	...	...	...	...	...
3. Iron foundries	...	...	...	...	...	...	...	...
V. STOCK VARIATION (total)	...	...	...	...	...	...	...	...
1. Iron and steel works	...	...	6	...	...	61	45	45
2. Steel foundries (independent)	...	...	...	...	...	...	...	...
VI. FOREIGN TRADE (balance)	...	...	-3 643	-6 359	-5 881	-6 839		...
1. Imports (-)	...	...	-3 650	-6 374	-5 917	-6 852	...	...
2. Exports (+)	...	...	7	15	36	13	16	...
A. TOTAL NATIONAL RESOURCES (utilized)(II + III + VI)	...	...	-2 901	...	...	-4 665	...	...
B. DOMESTIC SCRAP (total) (A - II.1)(collection utilized)	...	...	...	...	...	1 536	...	...
C. DOMESTIC SCRAP MARKETED (B - II.2) or (A - II)	...	...	...	...	...	-5 337	...	...
D. EXTERNAL PURCHASE REQUIREMENTS (GROSS)(I - II) (all consumers)	...	...	5 845	7 748	...	10 097	10 053	
E. EXTERNAL PURCHASE REQUIREMENTS (NET) (I - II + IV) (all consumers)	...	...	...	...	...	...	...	...
F. SURPLUS (+) or DEFICIT (-) (balance) (C - E) or (VI + V)	...	...	...	...	...	...	...	...

Source: Government replies to UN/ECE questionnaires.

UKRAINE

(Thousands of tonnes)

	1985	1990	1991	1992	1993	1994	1995	1996
Basic Iron and Steel Data								
1. Pig iron production	...	41 906	36 632	35 295	27 108	20 118	17 588	17 420
2. DRI production	...	...	...	...	...	...	...	...
3. Crude steel production	...	52 635	44 761	41 429	32 421	24 081	22 309	22 332
4. CC ratio (%)	...	8	8	8	6	8	16	20
Iron and Steel Scrap Data								
I. CONSUMPTION (total)	...	26 269	21 964	19 130	14 805	10 223	9 633	...
1. Iron and steel works	...	26 269	21 964	19 130	14 805	10 223	9 633	...
2. Steel foundries	...	...	...	...	...	...	...	...
3. Iron foundries	...	...	...	...	...	...	...	...
4. Re-rolling	...	...	...	...	...	...	...	...
5. Independent electric ferro-alloy	...	...	...	...	...	...	...	...
II. OWN RESOURCES (total)	...	14 738	12 618	11 292	9 578	5 746	5 378	...
1. Circulating scrap	...	13 442	11 479	10 310	7 548	5 139	4 894	...
(a) Iron and steel works	...	13 442	11 479	10 310	7 548	5 139	4 894	...
(b) Steel foundries (independent)	...	...	...	...	...	...	...	...
(c) Iron foundries	...	...	...	...	...	...	...	...
2. Capital scrap (demolition, etc.)	...	1 296	1 139	983	2 030	607	484	...
III. EXTERNAL RECEIPTS (tota	...	6 039	5 427	4 502	5 129	3 637	...	...
1. Iron and steel works	...	6 039	5 427	4 502	4 469	3 447	...	...
(a) Trade	...	...	...	...	...	...	...	...
(b) Direct purchase	...	...	...	...	...	...	...	...
(c) Imports (not included in (a) or (b	...	...	...	...	...	...	...	...
2. Steel foundries (independent)	...	...	...	...	...	...	...	...
3. Iron foundries	...	...	...	...	660	190	...	...
4. Independent electric ferro-alloy	...	...	...	...	...	...	...	...
IV. OUTSIDE SALES (-) (total)	...	...	...	...	...	...	...	...
1. Iron and steel works	...	...	...	...	...	...	...	...
2. Steel foundries (independent)	...	...	...	...	...	...	...	...
3. Iron foundries	...	...	...	...	...	...	...	...
V. STOCK VARIATION (total)	...	...	...	...	...	...	...	...
1. Iron and steel works	...	...	...	...	...	...	...	...
2. Steel foundries (independent)	...	...	...	...	...	...	...	...
VI. FOREIGN TRADE (balance)	...	...	...	230	107	716	13	620
1. Imports (-)	...	...	...	- 249	- 159	- 113	- 181	-80
2. Exports (+)	...	...	...	479	266	829	194	700
A. TOTAL NATIONAL RESOURCES (utilized)(II + III + VI)	...	...	...	16 024	14 814	10 099	...	...
B. DOMESTIC SCRAP (total) (A - II.1)(collection utilized)	...	...	...	5 715	7 266	4 960	...	...
C. DOMESTIC SCRAP MARKED (B - II.2) or (A - II)	...	...	...	4 732	5 236	4 353	...	...
D. EXTERNAL PURCHASE REQUIREMENTS (GROSS)(I - II) (all consumers)	...	11 531	9 346	7 838	5 227	4 477	4 255	...
E. EXTERNAL PURCHASE REQUIREMENTS (NET) (I - II + IV) (all consumers	...	...	...	...	...	...	...	...
F. SURPLUS (+) or DEFICIT (-) (balance) (C - E) or (VI + V)	...	...	...	...	...	...	...	...

Source: Government replies to UN/ECE questionnaires.

UNITED KINGDOM

(Thousands of tonnes)

	1980	1985	1990	1992	1993	1994	1995	1996
Basic Iron and Steel Data								
1. Pig iron production	6 412	10 458	12 463	11 542	11 534	11 943	12 236	12 830
2. DRI production	...	...	-	-	-	-	-	-
3. Crude steel production	11 278	15 722	17 601	16 212	16 625	17 286	17 423	17 818
4. CC ratio (%)	27	55	85	86	86	87	88	89
Iron and Steel Scrap Data								
I. CONSUMPTION (total)	9 297	...	8 865	6 190	6 550	6 874	7 196	7 019
1. Iron and steel works	6 765	...	7 016	6 019	6 379	6 699	7 012	6 830
2. Steel foundries	...	...	240	171	171	175	184	189
3. Iron foundries	2 416	...	1 609	...	...	...	...	...
4. Re-rolling	...	...	...	...	...	...	...	...
5. Independent electric ferro-alloy	...	...	...	...	...	...	...	...
II. OWN RESOURCES (total)	4 335	...	3 181	1 690	1 651	1 696	2 476	2 730
1. Circulating scrap	3 978	...	3 181	1 605	1 576	1 620	2 476	2 730
(a) Iron and steel works	3 130	...	2 375	1 605	1 576	1 620	2 390	2 640
(b) Steel foundries (independent)	...	...	113	...	...	...	86	90
(c) Iron foundries	848	...	693	...	...	...	...	...
2. Capital scrap (demolition, etc.)	357	...	...	85	75	76		
III. EXTERNAL RECEIPTS (total)	4 592	...	...	3 974	4 288	4 641	4 794	
1. Iron and steel works	2 988	...	...	3 974	4 288	4 641	...	...
(a) Trade	2 987	...	...	3 628	3 949	4 287	...	...
(b) Direct purchase	1	...	...	341	328	346	...	...
(c) Imports (not included in (a) or (b))	-	...	...	5	11	8	...	...
2. Steel foundries (independent)	...	...	...	...	...	...	...	...
3. Iron foundries	1 502	...	...	...	...	...	...	...
4. Independent electric ferro-alloy	...	...	...	-	-	-	...	...
IV. OUTSIDE SALES (-) (total)	102	...	...	25	33	23	...	...
1. Iron and steel works	102	...	...	25	33	23	...	...
2. Steel foundries (independent)	...	...	...	...	...	...	...	...
3. Iron foundries	...	...	...	...	...	...	...	...
V. STOCK VARIATION (total)	- 472	...	...	- 551	- 644	- 560	...	...
1. Iron and steel works	- 392	...	...	4	- 95	- 25	...	...
2. Steel foundries (independent)	...	...	...	...	...	...	...	...
VI. FOREIGN TRADE (balance)	2 781	...	3 138	2 727	3 819	3 452	3 266	3 172
1. Imports (-)	- 26	...	- 60	- 69	- 47	- 94	- 176	-247
2. Exports (+)	2 807	...	3 198	2 796	3 866	3 546	3 442	3 419
A. TOTAL NATIONAL RESOURCES (utilized)(II + III + VI)	11 708	...	...	8 391	9 758	9 789	10 536	...
B. DOMESTIC SCRAP (total) (A - II.1)(collection utilized)	7 730	...	...	6 786	8 182	8 169	...	...
C. DOMESTIC SCRAP MARKETED (B - II.2) or (A - II)	7 373	...	...	6 701	8 107	8 093	8 060	...
D. EXTERNAL PURCHASE REQUIREMENTS (GROSS)(I - II) (all consumers)	4 962	...	5 684	4 500	4 899	5 178	4 720	4 289
E. EXTERNAL PURCHASE REQUIREMENTS (NET) (I - II + IV) (all consumers)	5 064	...	...	4 525	4 932	5 201	...	...
F. SURPLUS (+) or DEFICIT (-) (balance) (C - E) or (VI + V)	2 309	...	...	2 176	3 175	2 892	...	...

Source: Government replies to UN/ECE questionnaires.

Note: Stock variation calculated as follows: V = II + III - I - IV.

UNITED STATES

(Thousands of tonnes)

	1980	1985	1990	1993	1994	1995	1996
Basic Iron and Steel Data							
1. Pig iron production	62 362	45 282	49 668	48 155	49 374	50 890	49 376
2. DRI production	...	203*	390	440	480	460	450
3. Crude steel production	101 456	80 068	89 726	88 793	91 243	95 191	94 670
4. CC ratio (%)	20	44	67	86	90	91	93
Iron and Steel Scrap Data							
I. CONSUMPTION (total)	76 201	62 978	56 013	67 334	55 800	71 655	53 700
1. Iron and steel works	60 380	47 677	46 196	52 947	54 000	56 655	53 700
2. Steel foundries	2 582	1 875	2 664	1 729	1 800	2 000	...
3. Iron foundries	12 965	13 426	7 153	12 658	...	13 000	...
4. Re-rolling	...	...	...	...	...	...	...
5. Independent electric ferro-alloy	...	...	...	...	...	...	...
II. OWN RESOURCES (total)	38 263	27 621	31 806	21 476	20 460	29 721	...
1. Circulating scrap	36 664	26 687	30 976	20 925	20 100	29 531	...
(a) Iron and steel works	31 941	21 414	25 144	15 858	15 000	20 000	...
(b) Steel foundries (independent)	819	498	756	565	...	630	...
(c) Iron foundries	3 904	4 775	5 076	4 502	5 100	8 901	...
2. Capital scrap (demolition, etc.)	1 599	934	830	551	360	190	...
III. EXTERNAL RECEIPTS (total)	42 608	35 214	52 612	46 294	49 400	59 650	...
1. Iron and steel works	31 325	24 951	40 430	37 675	41 000	50 000	...
(a) Trade	30 819	24 951	...	...	...	...	...
(b) Direct purchase	...	...	...	...	...	...	...
(c) Imports (not included in (a) or (b))	506	...	...	...	...	...	...
2. Steel foundries (independent)	1 750	1 453	2 184	1 143	...	1 300	...
3. Iron foundries	9 259	8 810	9 998	7 476	8 400	8 350	...
4. Independent electric ferro-alloy	...	...	...	...	...	...	...
IV. OUTSIDE SALES (-) (total)	5 336	...	...	...	...	...	...
1. Iron and steel works	4 951	...	...	...	...	...	...
2. Steel foundries (independent)	36	...	...	...	...	...	...
3. Iron foundries	349	...	...	...	...	...	...
V. STOCK VARIATION (total)	- 666	...	...	...	...	...	...
1. Iron and steel works	- 471	- 439	...	161	...	...	...
2. Steel foundries (independent)	- 49	76	...	- 75	...	- 2	...
VI. FOREIGN TRADE (balance)	9 625	8 481	10 295	8 324	6 962	8 343	5 981
1. Imports (-)	- 506	- 545	-1 285	-1 545	-1 877	-2 096	-2 564
2. Exports (+)	10 131	9 026	11 580	9 869	8 839	10 439	8 545
A. TOTAL NATIONAL RESOURCES (utilized)(II + III + VI)	90 496	71 316	94 713	76 094	76 822	97 714	...
B. DOMESTIC SCRAP (total) (A - II.1)(collection utilized)	53 832	44 629	63 737	55 169	56 722	68 183	...
C. DOMESTIC SCRAP MARKETED (B - II.2) or (A - II)	52 233	43 695	62 907	54 618	56 362	67 993	...
D. EXTERNAL PURCHASE REQUIREMENTS (GROSS)(I - II) (all consumers)	37 938	35 357	24 207	45 858	35 340	41 934	...
E. EXTERNAL PURCHASE REQUIREMENTS (NET) (I - II + IV) (all consumers)	43 274	...	...	...	...	...	...
F. SURPLUS (+) or DEFICIT (-) (balance) (C - E) or (VI + V)	8 959	...	...	...	...	...	...

Source: Government replies to UN/ECE questionnaires.

Notes: Stock variation calculated as follows: V = II + III - I - IV. Data for 1996 is preliminary.

ALBANIA

(Thousands of tonnes)

	1985	1990	1992	1993	1994	1995	1996	1997 Forecast
Iron and Steel Scrap Data								
VI. FOREIGN TRADE (balance)	...	...	...	...	...	...	...	...
1. Imports (-)	...	...	...	...	...	...	-55	...
2. Exports (+)	...	...	...	...	...	18710*	22001*	...

Source: Government replies to UN/ECE questionnaires.

Note: *Exports of scrap in millions of Albanian Lek.

AUSTRALIA

(Thousands of tonnes)

	1980	1985	1990	1993	1994	1995	1996	1997 Forecast
Basic Iron and Steel Data								
1. Pig iron production	6 976	5 599	6 127	7 414	7 467	7476	7 381	7 400
2. DRI production	...	...	...	...	...	-	-	-
3. Crude steel production	7 589	6 609	6 666	7 722	8 427	8494	8 392	8 400
4. CC ratio (%)	10	27	84	98	100	99	99	...
Iron and Steel Scrap Data								
VI. FOREIGN TRADE (balance)	...	...	1 300	757	349	308	670	...
1. Imports (-)	...	...	- 1	- 2	- 49	-2	-8	...
2. Exports (+)	...	...	1 301	759	398	310	678	...

Source: Government replies to UN/ECE questionnaires.

CROATIA

(Thousands of tonnes)

	1980	1985	1990	1993	1994	1995	1996	1997 Forecast
Basic Iron and Steel Data								
1. Pig Iron production	...	...	457	37	55	-	-	...
2. DRI production	...	...	...	...	...	-	-	-
3. Crude Steel production	...	...	424	74	63	45	46	
4. CC Ratio (%)	...	...	82	84	81	91	78	
Iron and Steel Scrap Data								
I. CONSUMPTION (total)	...	...	...	...	...	...	...	...
1. Iron and Steel Works	...	...	172	...	...	44	36	...
II. OWN RESOURCES (total)	...	...	...	...	...	...	...	...
1. Circulating scrap	...	...	...	...	...	...	...	...
(a) Iron and steel works	...	...	112	...	...	9	15	...
VI. FOREIGN TRADE (balance)	...	...	19	39	61	98	104	...
1. Imports (-)	...	...	-33	-6	-1	-2	-2	...
2. Exports (+)	...	...	52	45	62	100	106	...

Source: Government replies to UN/ECE questionnaires.

Note: Since 1992 the foreign trade data include foreign trade with the countries of former Yugoslavia.

CYPRUS

(Thousands of tonnes)

	1985	1990	1992	1993	1994	1995	1996	1997 Forecast
Iron and Steel Scrap Data								
VI. FOREIGN TRADE (balance)	...	14	17	19	22	26	34	35
1. Imports (-)	...	-1	-	-	-	-	-	-
2. Exports (+)	...	15	17	19	22	26	34	35

Source: Government replies to UN/ECE questionnaires.

ESTONIA

(Thousands of tonnes)

	1980	1985	1990	1993	1994	1995	1996	1997 Forecast
Basic Iron and Steel Data								
1. Pig iron production	...	...	...	-	-	3	3	...
2. DRI production	...	...	...	-	-	-	-	...
3. Crude steel production	...	...	...	-	-	3	3	...
4. CC ratio (%)	...	...	...	-	-	-	-	...
Iron and Steel Scrap Data								
VI. FOREIGN TRADE (balance)	...	...	...	202	190	251	186	...
1. Imports (-)	...	...	...	-1	-1	-5	-35	...
2. Exports (+)	...	...	...	203	191	256	221	...

Source: Government replies to UN/ECE questionnaires.

INDIA

(Thousands of tonnes)

	1980	1985	1989-90	1992-93	1993-94	1994-95	1995-96	1996-97
*Basic Iron and Steel Data**								
1. Pig iron production	8 509	9 840	11 930	15 126	15 647	17 808	19 025	19 864
2. DRI production	...	102	260	1 437	2 208	2 915	4 197	4 830
3. Crude steel production	9 514	11 936	14 609	18 117	18 155	19 282	20 768	21 795
4. CC ratio (%)	...	4	11	17	44	44	34	34
Iron and Steel Scrap Data								
VI. FOREIGN TRADE (balance)	...	...	-2 227	-2 573	- 754	-2 212	-1 966	-1 477
1. Imports (-)	...	...	-2 256	-2 573	- 754	-2 212	-1 966	-1 516
2. Exports (+)	...	...	29	...	...	-	-	39

Sources: Government replies to UN/ECE questionnaires; * International Iron and Steel Institute.

Notes: Pig-iron production data consist of hot-metal and cold pig-iron. Trade data for 1996-97 is for 11 months.

KAZAKSTAN

(Thousands of tonnes)

	1980	1985	1990	1993	1994	1995	1996	1997 Forecast
Basic Iron and Steel Data								
1. Pig iron production	...	...	5 226	3 544	2 435	2 592	2 536	...
2. DRI production	...	...	...	...	...	...	...	...
3. Crude steel production	...	...	6 753	4 554	2 969	3 027	3 216	...
4. CC ratio (%)	...	...	0	...	-	0	-	...
Iron and Steel Scrap Data								
VI. FOREIGN TRADE (balance)	...	...	...	...	...	152	144	...
1. Imports (-)	...	...	...	...	...	-1	-8	...
2. Exports (+)	...	...	...	...	...	153	152	...

Source: Government replies to UN/ECE questionnaires.

KYRGYZSTAN

(Thousands of tonnes)

	1980	1985	1990	1993	1994	1995	1996	1997 Forecast
Iron and Steel Scrap Data								
VI. FOREIGN TRADE (balance)	...	...	...	-19	-2	22	26	...
1. Imports (-)	...	...	...	-19	-3	-	-	...
2. Exports (+)	...	...	...	-	1	22	26	...

Source: Government replies to UN/ECE questionnaires.

LITHUANIA

(Thousands of tonnes)

	1980	1985	1990	1993	1994	1995	1996	1997 Forecast
Iron and Steel Scrap Data								
I. CONSUMPTION (total)	...	...	...	...	...	...	...	...
1. Iron and steel works	...	...	...	...	...	...	...	...
2. Steel foundries	...	...	...	...	...	...	...	...
3. Iron foundries	...	...	...	...	...	...	...	...
4. Re-rolling	...	...	124	...	...	150	123	...
5. Independent electric ferro-alloy	...	...	...	...	...	...	...	...
VI. FOREIGN TRADE (balance)	...	...	...	...	504	579	385	...
1. Imports (-)	...	...	...	...	-1	-7	-47	...
2. Exports (+)	...	...	...	...	505	586	432	...

Source: Government replies to UN/ECE questionnaires.

MALTA

(Thousands of tonnes)

	1990*	1992	1993	1994	1995	1995*	1996*	1997 Forecast
Iron and Steel Scrap Data								
VI. FOREIGN TRADE (balance)	513 601	9	...	5	3	585 490	428 763	...
1. Imports (-)	-19 017	-	-6	-		-61 530	-20 633	...
2. Exports (+)	532 618	9	...	5	4	647 020	449 396	...

Sources: Government replies to UN/ECE questionnaires.

Notes: *Foreign trade in LM (Maltese Liri). 1996 data for January to September.

REPUBLIC OF MOLDOVA

(Thousands of tonnes)

	1980	1985	1990	1993	1994	1995	1996	1997 Forecast
Basic Iron and Steel Data								
1. Pig iron production	...	...	...	...	...	...	...	...
2. DRI production	...	...	...	...	...	...	...	...
3. Crude steel production	...	...	712	497	-	0	0	...
4. CC ratio (%)	...	...	...	-	-	-	-	...
Iron and Steel Scrap Data								
VI. FOREIGN TRADE (balance)	...	...	...	...	3	9	2	...
1. Imports (-)	...	...	...	...	0	-1	-1	...
2. Exports (+)	...	...	...	...	3	10	3	...

Source: Government replies to UN/ECE questionnaires.

SLOVENIA

(Thousands of tonnes)

	1985	1990	1992	1993	1994	1995	1996	1997 Forecast
Basic Iron and Steel Data								
1. Pig iron production	...	-	-	-	-	-	-	...
2. DRI production	...	...	...	...	...	...	...	...
3. Crude steel production	...	528	320	372	442	427	348	...
4. CC ratio (%)	...	53	61	70	72	71	67	...
Iron and Steel Scrap Data								
I. CONSUMPTION (total)	...	606	391	338	416	394	301	...
1. Iron and steel works	...	606	391	338	416	394	301	...
VI. FOREIGN TRADE (balance)	...	-71	-253	-146	-244	- 212	- 88	...
1. Imports (-)	...	-148	-286	-196	-278	-260	-143	...
2. Exports (+)	...	77	33	50	34	48	55	...

Source: Government replies to UN/ECE questionnaires.

Note: For 1996 foreign trade data are provisional.

YUGOSLAVIA

(Thousands of tonnes)

	1980	1990	1992	1993	1994	1995	1996	1997 Forecast
Basic Iron and Steel Data								
1. Pig iron production	...	...	511	62	22	108	535	...
2. DRI production	...	...	...	...	...	...	...	...
3. Crude steel production	...	1 012	667	181	139	182	689	...
4. CC ratio (%)	...	...	86	75	70	86	95	...
Iron and Steel Scrap Data								
I. CONSUMPTION (total)	...	...	306	147	160	175	...	...
VI. FOREIGN TRADE (balance)	...	...	...	...	...	...	-75	...
1. Imports (-)	...	...	-105	...	...	...	-78	...
2. Exports (+)	...	...	...	...	...	...	3	...

Source: Government replies to UN/ECE questionnaires.

ARMENIA - No data available.

ICELAND - No production or trade in scrap from the Icelandic Alloy Ltd. No factories use scrap for an

ANNEX 2

SELECTED STATISTICAL TABLES

Table A2.1 Crude steel production by processes

(Percentage)

Country	Oxygen					Open-Hearth					Electric				
	1990	1993	1994	1995	1996	1990	1993	1994	1995	1996	1990	1993	1994	1995	1996
Australia	92	90	90	-	88	-	-	-	-	-	8	10	10	-	12
Austria	91	90	90	91	91	-	-	-	-	-	9	10	10	9	9
Belgium	91	87	86	85	86	-	-	-	-	-	9	13	14	15	14
Brazil	74	78	78	81	80	-	-	-	-	-	26	22	22	19	20
Bulgaria	53	58	67	65	65	8	-	-	-	-	42	42	33	35	35
Canada	63	64	61	61	62	-	-	-	-	-	40	36	39	39	38
Czech Republic	...	83	82	82	88	...	7	6	5	1	...	10	12	13	11
Finland	84	90	92	82	80	-	-	-	-	-	16	10	8	18	20
France	72	69	66	65	63	-	-	-	-	-	28	31	34	35	37
Germany *	81	77	78	76	74	...	-	-	-	-	...	23	22	24	26
Hungary	48	94	96	94	95	40	-	-	-	-	12	6	4	6	5
Italy	41	43	43	42	43	-	-	-	-	-	59	57	57	58	57
Japan	69	69	68	68	67	-	-	-	-	-	31	31	32	32	33
Kazakstan	70	...	...	85	78	23	...	...	12	19	7	...	...	3	3
Korea, Rep. of	69	67	64	62	61	-	-	-	-	-	31	33	36	38	39
Latvia	-	-	-	-	-	97	99	99	99	99	3	1	1	1	1
Luxembourg	100	100	85	54	47	-	-	-	-	-	-	-	15	46	53
Mexico	41	41	37	37	36	8	-	-	-	-	51	59	63	63	64
Netherlands	96	98	96	96	98	-	-	-	-	-	4	2	4	4	2
Poland	53	62	64	65	65	29	18	14	13	24	18	20	22	22	11
Portugal	47	54	61	54	54	-	-	-	-	-	53	46	39	46	46
Romania	54	54	57	62	63	22	18	17	15	15	24	28	26	23	22
Russian Federation	32	40	45	45	52	53	49	43	42	35	11	11	12	11	12
Spain	44	43	41	37	35	-	-	-	-	-	56	57	59	63	65
Slovakia	94	96	96	96	96	3	-	-	-	-	3	-	-	4	4
Sweden	61	64	64	63	65	-	-	-	-	-	39	36	36	37	35
South Africa	64	61	60	64	64	-	-	-	-	-	36	39	40	36	36
Turkey	40	30	30	29	32	6	6	6	5	5	54	64	64	66	63
United Kingdom	74	75	75	75	77	-	-	-	-	-	26	25	25	25	23
United States	59	60	60	60	58	4	-	-	-	-	37	40	40	40	42
Ukraine	40	41	45	43	44	53	52	50	51	50	7	7	7	6	6
European Union	69	64	64	-	-	-	-	-	-	-	31	36	36	-	-
World	55	57	56E	58	60	15	10	9E	-	-	28	31	33E	33	33

Sources: Government replies to UN/ECE questionnaires and figures of European Commission and IISI.
* Data for 1990 are for the former West Germany, as communicated.
** Data are for state-owned steel company SIDOR only.
Note: In Belarus, Croatia, Denmark, Estonia, Greece, Indonesia, Ireland, Lithuania, Republic of Moldova, Norway, Slovenia, Switzerland, Thailand and Venezuela, 100% of steel production is based on electric arc furnace as of the end of 1996.

Table A2.2 Continuous casting (percentage) and circulating scrap arisings in steelworks

(Kilograms per tonne and percentage)

	1985		1990		1995		1996	
	C C Ratio (%)	Circulating scrap (Kg/t)	C C Ratio (%)	Circulating scrap (Kg/t)	C C Ratio (%)	Circulating scrap (Kg/t)	C C Ratio (%)	Circulating scrap (Kg/t)
Australia	27	...	83	...	...	...	99	...
Austria	93	125	96	146	96	105	97	112
Belarus	...	...	...	66	...	86	...	93
Belgium	60	149	92	103	98	100	98	102
Brazil	44	...	58	158	...	...	...	...
Bulgaria	9	...	24	55	25	45	24	42
Canada	44	...	77	177	97	143	98	142
Czech Republic*	8	254	5	304	32	233	48	215
Finland	...	...	100	...	100	134	100	130
France	81	191	96	112	95	116	96	108
Germany**	80	174	91	120	95	112	96	113
Greece	...	...	100	106	100	100	100	112
Hungary	47	255	77	...	91	111	96	115
Ireland	...	140	100	...	100	29	100	38
Italy	79	114	95	107	97	81	97	83
Japan	91	...	94	93	96	96	96	102
Korea, Rep. of	...	...	96	85	98	72	98	69
Latvia	...	136	97	372	99	136	99	133
Luxembourg	28	162	34	137	46	176	53	154
Mexico	51	200	63	...	73	...	78	...
Netherlands	39	140	94	99	96	85	98	...
Norway	56	259	98	123	99	82	99	...
Poland	10	164	...	...	21	...	...	141
Portugal	43	...	54	240	100	340	96	141
Romania	...	226	36	...	48	308	49	223
Russian Federation***	4	...	25	307	37	195	41	103
Slovakia	...	226	26	153	89	112	100	...
Slovenia	...	...	53	...	71	...	67	...
South Africa	65	...	73	353	92	...	94	...
Spain	57	...	89	117	96	...	96	202
Sweden	81	...	86	219	86	216	88	...
Switzerland	66	...	100	59	100	...	100	63
Turkey	66	...	92	20	92	48	93	...
Ukraine	...	...	8	257	16	241	20	148
United Kingdom	55	...	85	86	88	137	89	...
United States	44	333	67	280	91	210	93	...
Venezuela ****	74	...	90	...	...	...	...	...

Sources: Government replies to UN/ECE questionnaires.

* Data for 1985 and 1990 are for the former Czechoslovakia, as communicated.

** Data for 1985 and 1990 are for the former West Germany, as communicated.

*** Data for 1985 are for the former Soviet Union, as communicated

**** Data are for state-owned steel company SIDOR only.

Table A2.3 Internal scrap arisings in iron foundries
(Thousands of tonnes and Kg/t of finished iron casting)

Country	1985		1990		1995		1996	
	Arisings	Kg/t	Arisings	Kg/t	Arisings	Kg/t	Arisings	Kg/t
France	653	270	715	270	709	270	683	270
Germany*	1343	290	1610	330	1590	357	...	-
Hungary	...	...	20	103	9	161	...	...
Japan	2262	...	3001	...	2617	...	2625	...
Korea, Rep. of	...	...	307	...	328	...	343	...
Luxembourg	6	111	2	69	-	-	-	-
Portugal	...	...	30	254	-	-	-	-
Russian Federation	...	...	2976	444	865	373	860	374
Slovakia	...	...	21	155	13	220	10	255
United States	4775	...	5076	...	8901	...	...	...

Sources: Government replies to UN/ECE questionnaires.
* Data for 1985 and 1990 are for the former West Germany, as communicated.

Table A2.4 Internal scrap arisings (other than circulating scrap) in own resources of steel works and steel foundries

(Thousands of tonnes and percentage)

Country	1985			1990			1995			1996		
	Own resources	Arising	%	Own resources	Arisings	%	Own resources	Arising	%	Own resources	Arisings	%
Austria	581	...	...	628	...	...	524	...	...	482	...	...
Belarus	...	...	...	84	11	13	76	12	16	96	14	15
Belgium	1872	276	15	1371	198	14	1311	155	12	1209	109	9
Brazil	3629	...	...	3248	...	...	...	...	...	...	...	...
Bulgaria	...	...	...	119	...	...	123	...	...	102	...	...
Canada	...	...	...	2152	...	...	2049	...	...	2075	...	...
Czech Republic*	4298	480	11	4120	500	12	1671	...	...	1550	...	...
Denmark	109	6	7	125	...	...	...	...	...	...	...	...
Estonia	...	...	...	...	...	...	430	5	1	434	4	1
Finland	...	...	...	429	...	...	430	4	1	434	5	1
France	3988	394	10	3233	290	9	3165	305	10	2926	284	10
Germany**	7845	798	10	6517	129	2	6600	166	3	...	182	...
Greece	120	...	...	106	...	...	103	10	10	95	0	0
Hungary	1066	149	14	779	21	2	256	50	20	242	27	11
Ireland	18	...	...	...	...	...	...	9	...	...	13	...
Italy	4084	738	18	2838	188	7	2350	68	3	2034	26	1
Japan	12015	...	...	13227	...	...	12393	...	...	12705	...	...
Korea, Rep. of	...	...	...	2267	...	...	2988	...	...	3022	...	...
Latvia	...	...	...	273	66	24	43	5	12	46	7	15
Luxembourg	746	210	28	669	180	27	544	79	15	470	80	17
Mexico	1202	...	...	1488	141	9	...	...	...	...	...	...
Netherlands	1102	...	...	534	29	5	543	...	...	...	...	...
Norway	130	...	...	108	61	56	56	15	27	...	...	...
Poland	5389	1209	22	7637	1743	...	...	...	...	...	...	...
Portugal	119	10	8	172	921	12	260	...	...	118	...	...
Romania	...	...	...	...	...	...	2251	229	10	...	...	...
Russian Federation ***	90400	60800	67	28798	3898	13	14833	3515	24	13940	3000	21
Spain	...	...	...	1492	...	...	...	...	...	...	...	...
Slovakia	...	...	...	799	48	6	463	...	...	392	...	...
Switzerland	...	...	...	65	...	...	...	...	...	...	...	...
Sweden	...	...	...	979	4	1	1141	73	6	1051	59	6
South Africa	...	...	...	3043	...	...	...	...	...	...	...	...
Turkey	342	36	11	881	141	16	672	33	5	895	48	5
United Kingdom	...	...	...	3181			2476	...	...	2730	...	...
United States	27622	934	3	25770	...	...	29721	190	1	...	...	...
Ukraine	...	...	...	14738	1 296	9	5378	...	...	...	...	...

Sources: Government replies to UN/ECE questionnaires

* Data for 1985 and 1990 are for the former Czechoslovakia, as communicated.

** Data for 1985 and 1990 are for the former West Germany, as communicated.

*** Data for 1985 are for the former Soviet Union, as communicated.

Table A2.5 Circulating scrap and capital scrap arisings in iron and steelworks and steel foundries
(Thousands of tonnes)

	1980	1985	1990	1993	1994	1995	1996
Austria	833	581	628	548	566	524	482
Belarus	...	...	73	107	70	64	82
Belgium	2548	1872	1371	1265	1357	1311	1209
Brazil	3221	3629	3248	3660	3479	...	...
Bulgaria	...	...	119	...	...	123	102
Canada	...	...	2152	2428	2157	2049	2075
Czech Republic	...	...	3068	...	...	1671	1550
Finland	...	...	429	472	439	430	434
France	4779	3335	2518	2347	2464	2456	2243
Germany*	7670	6502	4907	4908	5074	5010	...
Greece	...	...	106	86	98	103	95
Hungary	...	1066	759	358	348	247	...
Italy	5780	3719	2838	2344	2534	2350	2034
Japan	13264	9753	10225	9862	9925	9776	10080
Korea, Rep. of	...	...	1960	2497	2664	2660	2679
Latvia	...	...	273	...	...	43	46
Luxembourg	1109	740	667	579	601	544	470
Mexico	...	1202	...	...	...	...	...
Netherlands	966	1031	534	550	602	543	...
Norway	...	120	108	85	72	56	...
Poland	7208	5389	...	2853	3016	...	...
Portugal	...	119	172	129	118	260	118
Romania	...	...	...	2103	2325	2251	...
Russian Federation**	...	34900	25822	17018	14127	13080	14080
Slovakia	...	...	783	474	560	440	369
Spain	2148	...	1492	1514	1625	...	...
Sweden	1493	...	979	738	905	1141	1051
Switzerland	...	...	65	64	55	...	...
Turkey	...	342	639	758	779	895	651
Ukraine	...	...	14738	8678	5746	5378	...
United Kingdom	3487	...	1744	1651	1696	2476	2730
United States	34359	22846	25900	16423	...	20630	...

Sources: Government replies to UN/ECE questionnaires.
* Data for 1980, 1985 and 1990 are for the former West Germany, as communicated.
** Data for 1985 are for the former Soviet Union, as communicated.

Table A2.6 Other internal scrap arisings in iron and steelworks and independent steel foundries

(Thousands of tonnes)

	1980	1985	1990	1993	1994	1995	1997
Belarus	...	...	66	...	...	5	7
Belgium	421	276	198	84	161	155	109
Finland	...	...	...	...	...	4	5
France	700	394	290	379	290	305	284
Germany*	1391	798	129	284	229	166	182
Greece	...	...	...	...	4	10	0
Hungary	...	...	...	...	...	50	27
Italy	751	738	188	78	103	68	26
Latvia	...	...	66	...	...	5	7
Luxembourg	268	210	180	114	109	79	80
Norway	...	61	64	30	15	...	...
Poland	1743	1209	...	...	...	...	...
Portugal	...	10	21	23	9	...	...
Romania	...	...	...	605	539	229	...
Russian Federation**	...	3765	3656	2042	2184	3515	3000
Slovakia	...	...	48	22	88	0	0
Sweden	182	...	4	...	...	73	59
Switzerland	...	...	65	64	55	...	...
Turkey	...	36	11	11	...	33	48
Ukraine	...	...	1296	739	607	...	...
United Kingdom	355	...	118	75	76	...	...
United States	1599	934	...	551	...	190	...

Sources: Government replies to UN/ECE questionnaires.
* Data for 1980, 1985 and 1990 are for the former West Germany, as communicated.
** Data for 1985 are for the former Soviet Union, as communicated.

Table A2.7 Scrap arisings in iron foundries
(Thousands of tonnes)

	1980	1985	1990	1993	1994	1995	1996
France	835	653	715	612	681	709	683
Germany *	1760	1343	1610	1326	1435	1590	...
Hungary	...	...	20	8	9	9	...
Italy	640	635	...	...	...	...	...
Japan	2171	2262	3001	2512	2553	2617	2625
Korea, Rep. of	...	...	307	304	309	328	343
Luxembourg	2	6	...	...	...	...	...
Netherlands	77	71	...	...	...	...	...
Poland	429	...	...	...	...	...	...
Russian Federation	...	...	2976	1477	1079	865	860
Slovakia	...	...	21	15	14	13	12
Sweden	94	82	...	...	...	...	...
United States	3904	4775	5060	4502	...	8901	...

Sources: Government replies to UN/ECE questionnaires.
* Data for 1980, 1985 and 1990 are for the former West Germany, as communicated.

Table A2.8 Total scrap consumption
(Thousands of tonnes)

	1980	1985	1990	1993	1994	1995	1996
Austria	1694	1525	1236	1436	1442	1659	1511
Belarus	...	...	715	1007	911	636	859
Belgium	3814	3033	3065	2983	3337	3535	2950
Brazil	6468	6998	6574	6761	6554	...	...
Bulgaria	...	...	1742	1195	1402	1574	1319
Canada	...	7585	6554	7395	7303	7351	7154
Czech Republic*	7680	7107	5082	3299	3244	3352	3038
France	10368	8509	9029	8797	9694	10023	9826
Germany**	20093	18258	16672	17945	18712	20218	...
Greece	...	...	1095	1094	973	1070	964
Hungary	...	2050	1533	599	619	770	755
Indonesia	...	...	1366	...	...	...	...
Italy	19207	16380	16698	17434	17675	18182	15879
Japan	43237	43623	48204	43086	43074	43753	43866
Korea, Rep. of	...	...	11613	14708	16222	19164	20854
Latvia	...	...	457	310	350	315	334
Luxembourg	1577	1625	1387	1278	1510	1958	2011
Mexico	...	3487	2543	2732	...	4010	4451
Netherlands	1621	1504	955	1127	1187	1292	...
Norway	...	526	481	614	557	582	...
Poland	11864	8744	...	5413	5917	6377	...
Portugal	...	343	610	592	418	598	533
Romania	...	...	...	3265	3397	3519	...
Russian Federation***	...	90100	53842	31356	24469	27925	26903
Slovakia	...	...	1777	1284	1259	1195	1343
Slovenia	...	...	606	338	416	394	301
South Africa	...	...	3229	3514	2971	...	...
Spain	9062	10940	8920	9242	9771	10186	...
Sweden	2788	3313	2168	2282	2683	2466	2395
Switzerland	...	...	1215	1385	1210	...	...
Thailand	...	...	780	1112	1525	...	...
Turkey	...	...	6048	8485	8960	10769	10947
Ukraine	...	...	26269	14456	10153	9633	...
United Kingdom	9297	...	8865	6550	6873	7196	7019
United States	76201	62978	56013	67334	...	71655	53700
Venezuela****	...	...	712	699	702	...	...
European Union (12)	...	...	63929	...	...	77285	...

Sources: Government replies to UN/ECE questionnaires.

* Data for 1980, 1985 and 1990 are for the former Czechoslovakia, as communicated.

** Data for 1980, 1985 and 1990 are for the former West Germany, as communicated.

*** Data for 1985 are for the former Soviet Union, as communicated.

**** Data are for state-owned steel company SIDOR only.

Table A2.9 Total scrap consumption in all crude steel production including independent steel foundries

(Thousands of tonnes)

	1980	1985	1990	1993	1994	1995	1996
Austria	1694	1525	1236	1436	1442	1659	1511
Belarus	...	...	715	1007	911	636	859
Belgium	3710	3033	3065	2983	3337	3535	2950
Brazil	...	...	6574	6761	6554	...	...
Bulgaria	...	...	1742	1195	1402	1574	1319
Canada	...	7034	6554	7395	7303	7351	7154
Czech Republic	...	...	...	3299	3244	3352	3038
Denmark	811	595	608	...	...	...	...
France	8210	6486	7454	7116	7867	8035	7959
Germany*	15939	14480	12535	14548	14897	16007	...
Greece	...	...	1095	1094	973	1070	964
Hungary	...	1895	1394	548	573	726	...
Italy	17905	15403	16562	17368	17649	18171	...
Japan	36238	36713	40011	36409	36388	37004	37062
Korea, Rep. of	...	...	9764	13214	14638	17392	19201
Latvia	...	...	457	310	350	315	334
Luxembourg	1545	1571	1373	1278	1510	1958	2011
Mexico	...	3408	2279	2550	3495	4010	4451
Netherlands	1486	1350	955	1127	1187	1292	...
Norway	...	475	471	604	544	571	...
Poland	11037	8744	...	5202	5461	6377	...
Portugal	...	...	527	521	...	...	...
Romania	...	...	...	3265	3397	3519	...
Russian Federation**	...	77900	47516	28434	22370	26057	25133
Slovakia	...	...	1421	1094	1029	1150	1301
Slovenia	...	...	606	338	416	394	301
South Africa	...	...	3229	3514	2971	...	...
Spain	8528	10693	8920	9242	9771	10186	...
Sweden	2504	...	2168	2282	2683	2466	2395
Switzerland	...	...	1215	1385	1210	...	...
Thailand	...	...	780	1111	1524	...	...
Turkey	...	...	5760	8485	8960	10499	10502
Ukraine	...	...	26269	14456	10153	9633	...
United Kingdom	6765	...	7251	6550	6873	7196	7019
United States	62962	49552	48860	54676	...	58655	...
Venezuela***	...	...	712	690	702	...	...
European Union (12)	...	...	63929	...	...	77285	...

Sources: Government replies to UN/ECE questionnaires.

* Data for 1980, 1985 and 1990 are for the former West Germany, as communicated.

** Data for 1985 are for the former Soviet Union, as communicated.

*** Data are for state-owned steel company SIDOR only.

Table A2.10 Total circulating scrap consumption in iron and steelworks and independent steel foundries

(Thousands of Tonnes)

	1980	1985	1990	1993	1994	1995	1996
Austria	833	581	628	548	566	524	482
Azerbaijan	…	…	14	…	…	2	1
Belarus	…	…	73	107	70	64	82
Belgium	2127	1596	1173	1181	1196	1156	1100
Brazil	3221	3629	3248	3660	3479	…	…
Bulgaria	…	…	119	…	…	123	102
Canada	…	…	2152	2428	2157	2049	2075
Finland	…	…	429	472	439	426	429
France	4079	2940	2228	1968	2174	2151	1959
Germany*	6279	5704	4778	4624	4845	4844	…
Greece	…	…	106	86	93	94	95
Hungary	…	917	760	357	339	198	…
Italy	…	…	2650	2266	2431	2282	2008
Japan	13264	9753	10225	9862	9925	9776	10080
Korea, Rep. of	…	…	1960	2497	2664	2660	2679
Latvia	…	…	207	…	…	38	39
Luxembourg	841	530	487	465	492	460	386
Mexico	…	1202	1347	…	…	…	…
Netherlands	966	1031	534	550	602	543	…
Norway	…	120	47	45	42	41	…
Poland	…	4180	…	2853	3016	…	…
Portugal	…	109	151	106	109	…	…
Romania	…	…	…	1498	1786	2022	…
Russian Federation**	…	34900	22057	14229	12085	10453	10080
Slovakia	…	…	735	452	473	450	380
Spain	2148	…	1492	1514	1625	…	…
Sweden	1311	…	975	738	905	1068	992
Switzerland	…	…	65	64	55	…	…
Turkey	…	…	579	684	719	639	847
Ukraine	…	…	13442	7939	5139	…	…
United Kingdom	3130	…	2488	1576	1620	2476	2730
United States	32760	21912	25700	16423	…	20630	…

Sources: Government replies to UN/ECE questionnaires.

* Data for 1980, 1985 and 1990 are for the former West Germany, as communicated.

** Data for 1985 are for the former Soviet Union, as communicated.

Table A2.11 Specific consumption of scrap in total crude steel production
(Kilograms per tonne of steel produced in steelworks and independent steel foundries)

	1980	1985	1990	1993	1994	1995	1996
Austria	366	327	288	346	328	332	340
Belarus	...	...	966	1064	1043	855	970
Belgium	301	284	270	295	296	306	274
Brazil	...	342	320	268	255	...	...
Bulgaria	...	...	771	616	563	578	537
Canada	...	481	538	517	529	514	489
Czech Republic	...	...	...	487	458	467	467
France	354	345	392	416	436	450	457
Finland	...	...	...	...	...	358	342
Germany*	363	357	326	387	365	381	...
Greece	...	...	1096	1116	1147	1140	1138
Hungary	...	571	493	313	298	390	403
Ireland	...	...	...	...	...	1087	1141
Italy	676	645	666	675	675	657	656
Japan	325	349	363	365	370	364	375
Korea, Rep. of	...	...	401	397	434	473	494
Latvia	...	...	853	1033	1054	1129	1140
Luxembourg	334	398	386	388	491	749	804
Mexico	...	460	261	277	...	288	305
Netherlands	282	245	176	188	192	202	...
Norway	...	505	1104	1119	1121	1135	...
Poland	566	542	...	545	535	536	...
Portugal	...	...	735	694	578	751	626
Romania	...	...	...	600	586	537	...
Russian Federation	...	...	495	459	435	507	511
Slovakia	...	...	329	293	270	282	352
Slovenia	...	...	1148	909	941	923	865
South Africa	...	...	375	403	357	...	...
Spain	677	753	702	719	732	738	...
Sweden	591	...	487	497	541	498	488
Switzerland	...	...	1100	1104	1100	...	...
Thailand	...	...	1136	1143	1095	817	818
Turkey	...	...	618	743	740	432	...
Ukraine	...	...	499	446	424	413	394
United Kingdom	600	...	406	394	398	616	567
United States	621	544	605	616	...	...	...
Venezuela**	...	...	225	208	181	...	...
European Union (12)	...	...	467	...	...	496	...

Sources: Government replies to UN/ECE questionnaires.
* Data for 1980, 1985 and 1990 are for the former West Germany, as communicated.
** Data are for state-owned steel company SIDOR only.

Table A2.12 Total scrap consumption in blast furnaces

(Thousands of tonnes)

	1980	1985	1990	1993	1994	1995	1996
Austria	69	97	80	…	…	13	13
Belgium	9	…	3	0	19	46	49
Bulgaria	…	…	132	…	…	8	7
France	114	24	3	0	0	64	56
Germany*	475	374	115	115	73	51	46
Hungary	…	135	90	…	…	124	163
Italy	38	16	25	42	38	7	29
Japan	47	22	18	13	28	20	4
Luxembourg	3	2	…	…	…	…	…
Mexico	…	46	28	…	…	…	…
Poland	31	19	…	…	…	157	
Romania	…	…	…	394	342	334	320
Russian Federation**	…	1700	778	515	468	84	…
Spain	18	16	67	28	71	31	21
Sweden	50	2	31	27	6	…	…
Turkey	…	…	255	360	…	…	…
Ukraine	…	…	1851	375	239	…	…
United Kingdom	32	…	41	29	35	12	5
United States	3312	2251	2527	1929	…	1600	1700

Sources: Government replies to UN/ECE questionnaires.

* Data for 1980, 1985 and 1990 are for the former West Germany, as communicated.

** Data for 1985 are for the former Soviet Union, as communicated.

Table A2.13 Specific consumption of scrap in blast furnaces
(Kilograms per tonne of pig iron produced)

	1980	1985	1990	1993	1994	1995	1996
Austria	20	26	23	...	...	3	4
Bulgaria	...	....	116	...	...	...	...
Czech Republic*	14	13	...	...	...	...	...
France	6	2	...	...	...	5	5
Germany**	14	12	4	4	2	2	2
Hungary	...	34	53	...	...	82	109
Italy	3	1	2	4	3	1	3
Japan	0.6	0.3	...	0.2	0.4	0.3	0.05
Luxembourg	1	1	...	...	...	...	...
Mexico	...	13	8	...	...	...	...
Poland	3	2	...	...	...	...	...
Romania	...	...	...	123	98	37	...
Russian Federation	...	...	13	13	13	9	9
Spain	3	3	12	5	13	16	...
Sweden	21	1	11	9	2	10	7
Turkey	...	...	47	73	...	...	...
Ukraine	...	...	41	14	12	...	...
United Kingdom	5	...	3	2.5	3	1	0
United States	53	50	51	39	...	31	34

Sources: Government replies to UN/ECE questionnaires
* Data for 1980 and 1985 are for the former Czechoslovakia, as communicated.
** Data for 1980, 1985 and 1990 are for the former West Germany, as communicated.

Table A2.14 Total scrap consumption in oxygen converters
(Thousands of tonnes)

	1980	1985	1990	1993	1994	1995	1996
Austria	956	968	884	1053	1051	1174	1089
Belgium	2974	2223	2057	1708	1752	1828	1420
Brazil	…	…	4075	4789	4444	…	…
Bulgaria	…	…	235	…	…	491	441
Czech Republic*	1226	1834	2320	…	…	…	…
Finland	…	…	…	…	…	768	773
France	4183	2857	1856	1587	1389	1167	1052
Germany**	7227	6411	5499	5610	5915	5787	5321
Hungary	…	350	380	435	452	462	488
Italy	2419	1597	1165	1539	1919	1716	1388
Japan	7732	5565	5557	5384	5929	5555	5163
Korea, Rep. of	…	…	1875	2239	2562	2595	3026
Luxembourg	1542	1569	1373	1273	1022	614	474
Mexico	…	…	920	…	…	1311	1440
Netherlands	1366	1118	707	901	932	…	…
Poland	2190	1919	…	…	…	…	…
Portugal	…	75	41	83	87	88	87
Romania	…	…	…	674	758	1262	…
Russian Federation***	…	13100	7751	4896	4444	5116	4900
Slovakia	…	…	1347	1054	997	898	1021
South Africa	…	…	1000	1371	1114	…	…
Spain	1194	1214	986	989	981	866	…
Sweden	…	…	409	493	555	599	637
Turkey	…	…	558	553	634	603	747
Ukraine	…	…	5838	3133	2561	…	…
United Kingdom	1738	…	2291	1941	2203	…	…
United States	19830	13915	14713	14278	…	6000	15000

Sources: Government replies to UN/ECE questionnaires.

* Data for 1980, 1985 and 1990 are for the former Czechoslovakia, as communicated.

* Data for 1980, 1985 and 1990 are for the former West Germany, as communicated.

*** Data for 1985 are for the former Soviet Union, as communicated.

Table A2.15 Specific consumption of scrap in oxygen converters
(Kilograms per tonne of steel produced)

	1980	1985	1990	1993	1994	1995	1996
Austria	245	230	225	282	287	259	270
Belgium	255	226	198	193	181	185	154
Brazil	...	...	268	243	221	...	...
Bulgaria	...	...	241	...	...	278	264
Czech Republic*	311	297	...	...	...	...	...
Finland	...	...	...	...	...	372	255
France	221	189	136	133	117	101	96
Germany**	211	194	175	194	185	181	181
Hungary	...	...	263	263	248	264	274
Italy	202	140	...	...	...	146	133
Japan	92	74	73	79	88	81	78
Korea, Rep. of	...	...	118	101	118	113	128
Luxembourg	334	398	386	387	333	435	406
Mexico	...	210	242	...	...	261	289
Netherlands	270	211	136	155	157	...	...
Poland	288	281	...	...	...	...	...
Portugal	...	172	222	203	198	212	194
Romania	...	...	...	229	231	309	...
Russian Federation	...	...	273	222	210	209	193
Slovakia	...	...	307	280	262	237	298
South Africa	...	...	183	258	222	...	...
Spain	207	222	176	180	177	168	...
Sweden	165	151	151	168	175	193	200
Turkey	...	228	148	162	172	165	173
Ukraine	...	...	274	238	240	...	...
United Kingdom	260	...	174	157	171	...	...
United States	323	296	277	265	...	282	274

Sources: Government replies to UN/ECE questionnaires.
* Data for 1980 and 1985 are for the former Czechoslovakia, as communicated.
** Data for 1980, 1985 and 1990 are for the former West Germany, as communicated.

Table A2.16 Total scrap consumption in electric furnaces
(Thousands of tonnes)

	1980	1985	1990	1993	1994	1995	1996
Austria	557	460	272	383	391	472	409
Belarus	...	...	1074	1007	911	...	...
Belgium	663	810	1005	1276	1565	1661	1481
Brazil	...	...	2499	1972	2110	...	...
Bulgaria	...	...	935	...	...	1065	865
Croatia	...	...	172	...	...	44	36
Czech Republic*	1833	1981	1830	...	...	...	...
Denmark	611	595	655	...	...	...	...
Finland	...	...	...	...	...	310	272
France	3380	3597	5337	5338	6285	6607	6661
Germany**	5925	6035	6587	8551	8629	9885	10382
Greece	...	1064	1095	1094	973	1070	964
Hungary	...	422	225	100	97	110	92
Indonesia	...	...	1366	...	...	...	...
Italy	15051	13657	15246	15700	15592	16349	14461
Ireland	...	...	...	...	...	337	389
Japan	28458	31126	34436	31006	30427	31423	31891
Korea, Rep. of	...	...	7402	10975	12076	14797	16195
Latvia	...	...	14	4	4	4	3
Luxembourg	...	...	...	5	488	...	...
Mexico	...	...	1170	...	...	2184	2570
Netherlands	337	232	249	225	255	...	...
Norway	...	432	423	565	509	...	...
Poland	2570	2375	2447	2114	2445	...	...
Portugal	...	268	435	426	331	410	446
Romania	...	...	...	1402	1447	1187	...
Russian Federation***	...	7100	7751	4895	4444	5936	5680
Slovakia	...	...	0	0	0	210	239
Slovenia	...	...	606	338	416	394	301
South Africa	...	...	1986	1900	1686	...	...
Spain	7107	9459	7867	8225	8719	9236	...
Sweden	1917	2699	1728	1762	2122	1823	1699
Switzerland	...	...	215	1385	1210	...	...
Thailand	...	...	778	1111	1524	...	...
Turkey	...	...	5699	8375	8852	9776	9587
Ukraine	...	...	3379	2346	1417	...	...
United Kingdom	...	7100	4919	4580	4635	...	...
United States	29640	35043	32580	36140	...	39000	37000

Sources: Government replies to UN/ECE questionnaires.

* Data for 1980, 1985 and 1990 are for the former Czechoslovakia, as communicated.

** Data for 1980, 1985 and 1990 are for the former West Germany, as communicated.

*** Data for 1985 are for the former Soviet Union, as communicated.

Table A2.17 Specific consumption of scrap in electric furnaces
(Kilogram per tonne of steel produced)

	1980	1985	1990	1993	1994	1995	1996
Austria	969	1011	735	932	907	1040	1012
Belarus	...	...	966	1064	1043	...	...
Belgium	1056	1059	1019	989	971	989	962
Brazil	...	...	510	389	405	...	...
Bulgaria	...	...	1087	...	...	1132	1130
Croatia	...	...	1006	...	...	1073	1000
Czech Republic*	942	966	934	...	...	...	...
Finland	...	800	...	...	...	1023	1011
France	1010	1044	988	1023	1076	1047	1027
Germany**	982	1031	927	985	967	975	1003
Greece	...	1095	1096	1117	1148	1140	1138
Hungary	...	1094	1023	1590	1085	1145	1135
Ireland	...	...	...	...	...	1158	1142
Italy	1072	1090	1078	...	...	1026	1051
Japan	1045	1021	992	996	979	958	968
Korea, Rep. of	...	...	1030	1000	995	1065	1054
Latvia	...	...	833	1048	1032	1111	1239
Luxembourg	...	...	...	1250	1094	1117	1152
Mexico	416	413	261	...	...	288	305
Netherlands	1084	1089	1071	1196	1144	...	...
Norway	...	1061	1104	1119	1121	...	...
Poland	1047	975	...	...	...	...	...
Portugal	...	1117	1145	1246	1165	1165	1155
Romania	...	...	...	903	959	790	...
Russian Federation	...	...	860	834	834	886	927
Slovakia	...	...	...	...	...	1409	1513
Slovenia	...	...	1148	909	941	923	865
South Africa	...	...	654	567	523	...	...
Spain	1093	1083	1095	1115	1115	1068	...
Sweden	1019	1130	991	1067	1187	984	994
Switzerland	...	...	1099	1143	1095	...	...
Thailand	...	...	1136	1143	1095	...	...
Turkey	...	...	1150	1150	1153	1150	1150
Ukraine	...	...	938	975	1042	...	...
United Kingdom	1052	...	1053	1066	1059	...	...
United States	1048	1040	989	1034	...	1013	929
Venezuela***	...	...	299	272	262	...	...

Sources: Government replies to UN/ECE questionnaires.
* Data for 1980, 1985 and 1990 are for the former Czechoslovakia, as communicated.
** Data for 1980, 1985 and 1990 are for the former West Germany, as communicated.
*** Data are for state-owned steel company SIDOR only.

Table A2.18 Total scrap consumption in open-hearth furnaces

(Thousands of tonnes)

	1980	1985	1990	1993	1994	1995	1996
Azerbaijan	...	...	453	...	...	23	124
Hungary	...	...	760	-	-	-	
Latvia	...	...	417	306	346	310	330
Romania	...	...	...	722	748	676	...
Russian Federation*	...	40800	23061	13693	10098	10419	9980
Turkey	...	...	117	164	155	120	168
Ukraine	...	...	13705	7700	5491	...	
United States	...	...	1344	...	...	...	-

Sources: Government replies to UN/ECE questionnaires.

* Data for 1985 are for the former Soviet Union, as communicated.

Table A2.19 Specific consumption of scrap in open-hearth furnaces

(Kilograms per tonne of steel produced)

	1980	1985	1990	1993	1994	1995	1996
Azerbaijan	...	...	688	...	...	1200	829
Hungary	...	600	644	...	...	-	
Latvia	...	...	771	1 032	1 055	1123	1138
Mexico	...	454	395	...	...	...	-
Poland	700	633	...	...	...	...	
Romania	...	...	...	755	741	698	
Russian Federation*	...	475	521	513	501	514	565
Turkey	...	...	193	227	215	194	235
Ukraine	...	...	495	457	462	...	
United States	567	426	424	-	-	-	-

Sources: Government replies to UN/ECE questionnaires.

* Data for 1985 are for the former Soviet Union, as communicated.

Table A2.20 Total scrap consumption in independent steel foundries
(Thousands of tonnes)

	1980	1985	1990	1993	1994	1995	1996
Finland	...	...	...	...	...	16	13
France	309	208	250	191	193	197	190
Germany*	282	363	334	272	280	284	...
Hungary	...	...	29	13	14	...	...
Italy	157	133	126	87	100	99	...
Norway	...	4	48	39	35	...	...
Portugal	...	...	17	12	...	...	...
Russian Federation**	...	11900	7203	3241	2 063	1770	1770
Slovakia	...	...	97	40	32	42	41
United Kingdom	...	...	...	...	...	184	189
United States	2582	1875	2945	1729	...	2000	...

Sources: Government replies to UN/ECE questionnaires.
* Data for 1980, 1985 and 1990 are for the former West Germany, as communicated.
* Data for 1985 are for the former Soviet Union, as communicated.

Table A2.21 Total scrap consumption in iron foundries
(Thousands of tonnes)

	1980	1985	1990	1993	1994	1995	1996
France	2044	1695	1583	1681	1827	1899	1774
Germany*	4154	3778	4137	3397	3815	4211	...
Hungary	...	...	139	51	46	44	...
Italy	1200	810	...	...	...	...	...
Japan	6207	6400	7634	6293	6330	6438	6498
Korea, Rep. of	...	...	1425	1219	1298	1418	1422
Luxembourg	32	54	14	...	...	...	...
Netherlands	156	154	...	...	...	...	...
Portugal	...	...	83	71	...	...	...
Russian Federation**	...	11100	6013	2761	2015	1783	1770
Slovakia	...	...	73	64	51	29	22
Sweden	284	247	...	...	...	...	...
United States	12965	13426	15587	12658	...	13000	...

Sources: Government replies to UN/ECE questionnaires.
* Data for 1980, 1985 and 1990 are for the former West Germany, as communicated.
** Data for 1985 are for the former Soviet Union, as communicated.

Table A2.22 Specific consumption of scrap in iron foundries

(Kilograms per tonne of cast iron produced)

	1980	1985	1990	1993	1994	1995	1996
France	836	701	598	742	724	723	701
Germany*	...	815	848	851	876	887	...
Hungary	...	...	735	755	775	782	...
Italy	571	618	...	...	...	...	...
Luxembourg	582	482	483	-	-	-	-
Portugal	...	...	601	602	...	...	...
Russian Federation	...	...	898	973	973	770	...
Slovakia	...	...	696	764	720	491	468

Sources: Government replies to UN/ECE questionnaires.
* Data for 1985 and 1990 are for the former West Germany, as communicated.

Table A2.23 Total scrap consumption in other sectors (rerolling and electrometallurgy)

(Thousands of tonnes)

	1980	1985	1990	1993	1994	1995	1996	1997
France	114	128	91	93	96	89	93	...
Italy	102	167	136	66	26	11**	1**	...
Japan	792	510	559	384	356	311**	306**	...
Korea, Rep. of	...	...	424	275	286	354**	231**	200**
Mexico	...	...	264	182	...	...	...	...
Norway	...	...	10	10	13	...	...	...
Russian Federation*	...	1 100	313	161	84	85***	...	...
Slovakia	...	...	265	126	175	16	20	20

Sources: Government replies to UN/ECE questionnaires.
* Data for 1985 are for the former Soviet Union, as communicated.
** Without electrometallurgy.
*** Without rerolling.

Table A2.24 Total gross external requirements, all consumers
(Thousands of tonnes)

	1980	1985	1990	1993	1994	1995	1996
Austria	861	944	608	888	876	1135	1029
Azerbaijan	...	...	459	...	...	23	0
Belgium	1226	1161	1694	1718	1980	2224	1741
Brazil	3247	3369	3326	3101	3075	...	...
Bulgaria	...	...	1623	...	...	1451	1217
Canada	...	...	4402	4967	5146	5302	5079
Czech Republic	...	...	...	...	...	1681	1489
France	4754	4521	5895	5838	6549	6858	6900
Germany *	10663	10413	10155	11711	12203	131618	...
Greece	...	...	988	1008	875	967	869
Hungary	...	984	754	233	272	514	513
Italy	13287	12296	13860	15090	15141	15832	13845
Japan	27802	31608	34978	30712	30596	31360	31161
Korea, Rep. of	...	...	9346	11907	13249	16176	17832
Latvia	...	...	184	...	...	272	288
Luxembourg	466	879	720	699	909	1414	1541
Netherlands	578	402	421	577	585	...	...
Norway	...	396	373	529	485	526	...
Poland	4227	3355	...	2560	2901	...	...
Portugal	...	224	438	463	300	338	415
Romania	...	...	...	1162	1072	1268	...
Russian Federation	...	...	25044	12861	9263	13092	12963
Slovakia	...	...	978	795	684	732	951
South Africa	...	...	186	400	314	...	...
Spain	6864	...	7428	7728	8146	...	...
Sweden	1201	...	1189	1544	1542	1325	1344
Switzerland	...	...	1150	1321	1155	...	...
Turkey	...	...	5845	8210	8692	10097	10053
Ukraine	...	...	11531	5778	4407	4255	...
United Kingdom	4962	...	5684	4899	5177	4720	4289
United States	37938	35357	45126	45858	35340	41934	...
Venezuela **	...	...	36	473	388	...	...
European Union	...	...	49581	...	...	62023	...

Sources: Government replies to UN/ECE questionnaires.

* Data for 1980, 1985 and 1990 are for the former West Germany, as communicated.

** Data are for state-owned steel company SIDOR only.

Table A2.25 Total gross external requirements

(Percentage of total consumption)

	1980	1985	1990	1993	1994	1995	1996
Austria	50.8	61.9	49.2	61.8	60.7	68.5	68.1
Belarus	…	…	96.4	89.4	92.3	88.0	88.7
Belgium	32.4	38.3	55.2	57.6	59.3	62.9	59.0
Brazil	50.2	48.1	50.6	45.9	46.9	…	
Bulgaria	…	…	93.2	…	…	92.2	92.3
Canada			67.2	67.2	70.5	72.1	71.0
Czech Republic*	32.1	36.4	39.8	…	…	50.1	49.0
France	45.8	53.1	64.6	66.4	67.6	68.4	70.2
Finland	…			…	…	62.1	61.5
Germany**	53.1	57.0	60.9	65.3	65.2	67.4	…
Greece	…	88.7	90.2	92.1	89.9	90.4	90.1
Hungary	…	48.0	49.1	38.9	43.9	66.7	67.9
Italy	69.2	75.1	83.0	86.6	85.7	87.1	87.2
Ireland	…		…	…	…	97.3	96.6
Japan	64.3	72.5	72.5	71.3	71.0	71.7	71.0
Korea, Rep. of	…	…	79.6	81.0	81.7	84.4	85.5
Latvia	…	…	42.5	…	…	86.3	86.2
Luxembourg	29.5	54.1	51.9	54.7	60.2	72.2	76.6
Mexico	…	65.5	38.4	…	…	…	
Netherlands	35.7	26.7	44.1	51.2	49.3	…	
Norway	…	75.3	77.5	86.2	87.1	90.4	…
Poland	35.6	38.4	…	47.3	49.0	…	…
Portugal	…	65.3	71.8	78.2	71.8	56.8	77.9
Romania	…		…	35.6	31.5	36.0	
Russian Federation	…		54.2	52.6	55.5	46.9	48.2
Slovakia	…	…	55.2	61.9	54.3	61.2	70.8
South Africa	…	…	5.8	11.4	10.6	…	
Spain	75.7		83.3	83.6	83.4	…	…
Sweden	43.1	…	54.8	67.7	66.3	53.7	56.1
Switzerland	…	…	94.7	95.4	95.5	…	
Turkey	…	…	96.4	96.8	97.0	93.7	91.8
Ukraine		…	…	40.0	43.4	44.1	…
United Kingdom	53.4	…	75.9	74.8	75.3	65.6	61.1
United States	43.8	56.1	64.6	68.1	63.3	58.5	…
European Union	…	…	77.6			80.2	…

Sources: Government replies to UN/ECE questionnaires.

* Data for 1980, 1985 and 1990 are for the former Czechoslovakia, as communicated.

** Data for 1980, 1985 and 1990 are for the former West Germany, as communicated.

Table A2.26 Total net external requirements, all consumers
(Thousands of tonnes)

	1980	1985	1990	1993	1994	1995	1996
Austria	861	944	608	888	876	1139	1034
Belgium	1488	1303	1754	1751	2015	2257	1770
Brazil	3247	3369	3326	3101	3075	...	...
France	4985	4729	5895	5838	6569	6878	6928
Finland	...	...	...	...	...	722	714
Germany*	10893	10903	10675	12270	12869	14227	...
Greece	...	...	1012	1023	895	971	873
Hungary	...	1060	754	233	272	514	513
Italy	13759	12741	14094	15147	15195	15906	13922
Ireland	...	...	...	...	...	328	376
Japan	31447	34392	...	...	...	...	...
Korea, Rep. of	...	...	9384	...	...	16226	17844
Latvia	...	...	372	...	...	...	...
Luxembourg	599	931	743	751	998	1480	1608
Netherlands	578	402	421	...	...	...	...
Norway	...	396	373	529	485	526	...
Poland	4951	3624	...	...	3128	...	...
Portugal	...	225	438	463	300	338	415
Romania	...	...	...	1741	1734	1753	...
Russian Federation	...	...	29030	14675	10383	19374	19263
Slovakia	...	...	978	795	684	732	951
Spain	6864	...	7564	7856	8299	...	...
Sweden	1214	...	1189	1544	1783	1349	1388
Switzerland	...	...	1150	1321	1155	...	...
United Kingdom	5064	...	5547	4932	5200	...	...
United States	43274	35357	45126	45858	...	...	...

Sources: Government replies to UN/ECE questionnaires.

* Data for 1980, 1985 and 1990 are for the former West Germany, as communicated.

Table A2.27 Foreign trade balance and gross external requirements
(Percentage)

	1980	1985	1990	1993	1994	1995	1996
Austria	15	22	-47.4	-4.8	-6.7	3.5	...
Belarus	...	...	...	-28.4	-21.3	-3.6	29.3
Belgium	-7.4	64.9	60.0	49.3	27.0	79.5	118.9
Brazil	0.6	3.1	3.4	3.2	4.8	...	...
Canada	...	...	-6.8	3.3	2.0	-9.2	-15.8
Czech Republic*	7.8	-4.9	-5.2	...	...	-41.6	-56.3
Finland	...	...	...	...	...	16.7	-1.0
France	-60.1	-77.0	-43.0	-43.0	-33.3	-30.0	-35.7
Germany**	-14.7	-17.3	-39.4	-65.0	...	-49.7	...
Greece	...	...	101.8	34.9	13.2	...	...
Hungary	...	-4.5	3.0	-36.5	-192.6	-97.1	-114.4
Ireland	...	...	...	40.3	...	53.6	47.3
Italy	55.7	44.4	41.8	39.0	40.6	39.8	32.5
Japan	9.8	9.8	1.9	-0.8	0.3	0.9	-5.2
Korea, Rep. of	...	...	43.4	41.1	38.3	30.9	28.6
Luxembourg	88.6	92.9	87.4	90.6	101.0	97.4	98.7
Mexico	...	29.0	25.3	...	...	...	...
Netherlands	-181	-311	-343	...	...	...	...
Norway	...	1.5	...	40.3	28.2	25.5	...
Poland	3.6	0.1	...	-39.5	-15.5	...	...
Portugal	...	27.7	15.5	-8.0	-2.9	-9.1	-5.5
Romania	...	...	...	-1.5	-0.5	-0.6	...
Russian Federation	...	...	...	-3.0	-2.4	...	...
Slovakia	...	...	...	-21.9	-23.7	5.2	-5.6
South Africa	...	...	-38.2	-20.0	-24.8	...	...
Spain	58.6	...	57.4	51.5	56.6	...	...
Sweden	4.9	...	10.5	15.3	17.5	1.4	0.3
Switzerland	...	...	6.3	27.4	7.5	...	...
Turkey	...	...	65.6	77.4	69.8	67.7	...
Ukraine	...	...	...	-1.9	-16.2	-0.3	...
United Kingdom	-56.0	...	-57.0	-76.6	-66.7	-69.2	-74.5
United States	-25.3	-24.0	-22.7	-18.0	-19.7	-19.9	...
European Union	...	...	7.4	...	...	-3.0	...

Sources: Government replies to UN/ECE questionnaires.
Note: Figures in negative values mean that the country is a net exporter.
* Data for 1980, 1985 and 1990 are for the former Czechoslovakia, as communicated.
** Data for 1980, 1985 and 1990 are for the West Germany, as communicated.